Committed

Committed

Dispatches from a Psychiatrist in Training

Adam Stern, MD

Houghton Mifflin Harcourt
Boston New York
2021

For information about permission to reproduce selections from this book, write to trade.permissions@hmhco.com or to Permissions, Houghton Mifflin Harcourt Publishing Company, 3 Park Avenue, 19th Floor, New York, New York 10016.

hmhbooks.com

Library of Congress Cataloging-in-Publication Data
Names: Stern, Adam (Psychiatrist), author.
Title: Committed : dispatches from a psychiatrist in training / Adam Stern.
Description: Boston : Houghton Mifflin Harcourt, 2021.
Identifiers: LCCN 2020057716 (print) | LCCN 2020057717 (ebook) | ISBN 9780358434733 (hardcover) | ISBN 9780358450351 | ISBN 9780358450511 | ISBN 9780358435488 (ebook)
Subjects: LCSH: Stern, Adam (Psychiatrist) | Harvard Medical School. | Psychiatrists — United States — Biography. | Residents (Medicine) — United States — Biography. | Psychiatry — Study and teaching (Residency) — United States.
Classification: LCC RC438.6.S75 A3 2021 (print) | LCC RC438.6.S75 (ebook) | DDC 616.89092 [B] — dc23
LC record available at https://lccn.loc.gov/2020057716
LC ebook record available at https://lccn.loc.gov/2020057717

Book design by Chloe Foster

Printed in the United States of America
1 2021
4500826709

For my Golden Classmates,

our teachers,

and our patients

Author's Note

This book is based upon my experiences during psychiatry residency training. When possible, I have obtained permission from, and I have also changed names and identifying characteristics of, many of the people represented to protect their privacy. I've also created composite characters. Patient descriptions and patient encounters in this book have been deliberately altered to make them unidentifiable.

Contents

Prologue

I have a recurring dream in which I look down and notice for the first time that I'm soaring above the earth. I'm exhilarated but also filled with fear. I don't know how I made it off the ground, and the act of looking down seems to cause me to lose whatever momentum it was that propelled me upward. I need to figure out how to keep moving before gravity pulls me back to earth, ending in a terrible crash. Sometimes I awaken just as I begin to fall, and other times the dream ends with my discovery of an unexpected solution. The version that gives me the most comfort is when I look to one side or the other and notice that I'm not alone. In those moments, when I see someone floating right next to me, my fear still exists, but it's more surmountable. Maybe we can figure this out together.

In my conscious life I feel this way too. I became a psychiatrist, which was, at its core, an education in the value of human connection. Psychiatry is the field of medicine aimed at helping patients to find and become the best versions of themselves in spite of, or even because of, the immense challenges they face. Inherent within the field is the assumption that we're more capable together than we are apart.

Psychiatrists are trained to give people a push forward when they're stagnating and to catch them when they're falling. We learn through experience that the parts of our lives we can see and hear and feel may be only a sliver of our inner world. In fact, our minds are generally so focused on the complex task of making sense of our precarious existence that sometimes we can misunderstand parts of ourselves that exist right out in the open. Here, too, psychiatry can be useful in illuminating the unseen.

Those of us who become psychiatrists face these same challenges in our own lives. I chose it as my specialty because I wanted to become an expert in the human condition, but I had to figure out how to square that with the gnawing sense that sometimes I could just barely get by myself. I couldn't fathom how I might grow into an intellectual and emotional guide for patients when I felt completely overwhelmed by my own rudimentary life.

Coming from a state medical school in upstate New York, I had matched into a residency program at Harvard Medical School. Everyone around me at the new program was so bright and already accomplished that I didn't see myself belonging. I felt like an impostor.

A version of my dream was playing out in my life then, but I couldn't see how it would end. I had found myself soaring into one of the most prestigious residency programs in the country, but I couldn't imagine a scenario that didn't involve crashing back to earth.

This book tells the story of how I was transformed over those four years alongside my peers with the guidance of our extraordinary teachers. It describes how we all came together to overcome the unimaginable challenges of psychiatric training. Like many of the patients who taught us by way of their own care, my classmates and I changed and strove to be better. Together we learned the meaning of failure and appreciated the preciousness of success. As a group thrust together and inseparable by circumstance, our class was taught what it meant to connect with one another and our patients. We faltered often, but still, and always, we found ways to move forward together.

Part 1

Year One

1

Welcome to Longwood

The room was dimly lit. The curtains hung with too much slack, letting in light from the streetlamps outside the window. We stood in a patient's room at the far end of the locked psychiatry unit on the Longwood medical campus. I held steady in the center of the room hoping that the adrenaline rushing through my veins was evident only to me. I was flanked by three hospital security guards. One seemed to yawn in my ear as I stared up at a man, no more than twenty, who was admitted to the unit in a psychotic state. His world did not conform to reality, and his instincts or terrified thoughts had led him to climb to the top of a six-foot bureau. He perched completely frozen, crouching in fear.

"Come on down. We're only here to help," I said in a quiet voice.

"You're an agent. You're an agent of the Devil's CIA," he replied.

"Please," I pleaded even more soothingly. "I need you to come dow—" But before I finished my sentence, he had leapt toward us.

Two of the security guards intercepted him midair and guided him to the floor safely but with a thud. A nurse entered and proceeded to inject his buttocks with a sedative as the guards held him in place. She'd had it ready from the moment she walked in.

"I'm so sorry this has happened," I said, kneeling on the floor and trying to make eye contact with the man. "We're going to get through this together."

The group escorted the man to the Quiet Room, where he was physically tethered to the bedpost in accordance with Massachusetts legal code and the murky ethics of involuntary treatment when there is an imminent safety risk. I watched the man be tied down from eight feet away.

I felt a nudge in my rib.

"First time?" It was the nurse. "It gets easier."

"I'm not sure that I want it to get easier," I replied.

"Well, now, that's your stuff getting in the way." She sighed and slumped her shoulders. "Come on, *Doctor.* We've got mounds of paperwork and three admissions waiting to be seen."

One Month Earlier

"Have you ever seen anything like this?" I asked like an amazed child.

Eliana paused to take in the immaculate surroundings and shook her head.

"I don't think there's any place quite like this," she replied, fixing her gaze upon my Harvard-issued ID badge.

"Seeing that name next to mine feels like a lot of pressure," I said.

"Good thing you're *wicked smaht,*" she retorted with an exaggerated Boston accent.

I smiled meekly and continued examining the main quad at Harvard Medical School.

"This quad belongs to people way smarter than I am."

The locals sometimes refer to the school's campus and surrounding buildings as a "medical mecca," with esteemed institutions such as Beth Israel Deaconess Medical Center, Brigham and Women's Hospital, Boston Children's Hospital, and Dana Farber Cancer Institute, among others, all crammed into a five-block radius. It is the epicenter for discovery and execution in bioscience. Many top physician-scientists come through its grounds at one point or another. The gravitas is inescapable. In fact, the entire quad is mostly encased in marble, from the walls to the guardrails to the staircases.

Walking through that quad in June with my newly ex-girlfriend was probably a mistake. After being together for four years across college and medical school, Eliana and I still shared the rhythms and comforts of devoted partners. It was easy to forget that we weren't actually a pair anymore. We had broken up a few times over the prior year, but my matching at Harvard for residency training while her life was just taking off in New York was the final blow for our relationship. It was generous of her to visit me as I got settled into my new home, but her presence also highlighted the overwhelming uncertainty of starting anew by myself at a place as daunting as Harvard Medical School.

I had never felt smaller or more fragile in my life than I did walking through that quad for the first time. The crumbling brick and discolored concrete I had grown comfortable with at my state medical school were nowhere to be seen. I would only later learn that there were crumbled bricks here at Harvard too, just hidden away in the older wings of buildings that not coincidentally housed the financially challenged psychiatric services of the medical center.

It was my third day in Boston, but I had yet to sleep a full night. My nerves were fried. I didn't know anyone. I didn't even know what it meant to be a Harvard psychiatrist. I couldn't shake the idea that the Match algorithm that decides where med students will spend their residency training years had made a terrible mistake. I was coming from the State University of New York's Upstate Medical University — a wonderful place to study, in Syracuse — but also a school few in Boston even knew about, which I assumed often led to lower Match rankings by residency programs. My medical school's name — Upstate — was so generic that some people thought I had made it up.

My classmates-to-be, on the other hand, were generally entering our psychiatry residency from medical schools such as Yale, Duke, or Harvard itself — the same medical schools that hadn't even extended me the courtesy of an interview four years earlier.

Until medical school, I had always assumed that I was smart enough for most parts of life. I generally aced standardized tests and made it through my early education with ease. In the great lottery we all play at conception, I lucked out in many ways, not the least of which was that my intuitive ability in math perfectly maxes out at the exact difficulty level of the SATs. The hardest question on that test was the hardest question my brain was capable of solving, and that's one lucky coincidence. I worked hard to strengthen my college applications with a stellar GPA and several leadership positions in extracurricular programs.

As far as colleges were concerned, I was as smart as the smartest person in the world, though I knew very plainly that there were many, many people a lot smarter than I, who our society needed to do great, wonderful things that I could never achieve. I would never be a star mathematician or physicist solving the mysteries of the universe, but maybe I could do good for the people in it by being a doctor.

Becoming a doctor was a kind of family rite of passage for the Sterns. My father grew up with doctors in the family, and he decided to pursue medicine after attending MIT, thinking he might train to be a biomedical engineer, designing prosthetic devices. But the call of a stable job that rewarded his scientific mind was too much of a draw, and he went on to become a cardiologist. My older brother and I grew up admiring our father's ability to amply provide for us and be a positive force in the world simultaneously. A career in medicine was one of the paths we knew best that could reward our innate abilities in science. To be honest, if I thought I could have made it as a screenwriter in Hollywood, maybe this would be a very different memoir, but I've always been risk averse and choosing a career was no different. After performing below his natural ability in college, my brother took a few years afterward to buff up his application before getting into medical school. I learned from his experience, and in college I became a machine designed with one purpose —get into med school on the first attempt. And it worked. On September 15, 2005, the first day admissions were offered, I received a phone call from the admissions department at SUNY's Upstate Medical University telling me the good news. Soon after, my brother was admitted to SUNY Downstate College of Medicine in Brooklyn. It was a joyous time for our family.

Once we began medical school the following fall, my mood changed. The kind of intelligence required to thrive in medical school was something very different from what I had used my entire life. To do well, a medical student has to give up most of his life and spend nearly every waking minute studying. Devoting myself to the textbooks did not come naturally to me, and I quickly learned that there were dozens of people in my class with gifts for medical learning that I could only fantasize about. I watched them with awe as they applied rote memorization skills to the vast volumes of minutiae that enables human life — processes such as the Krebs cycle in biochemistry, which allows energy to be unleashed from aerobic respiration, or the elaborate cascade of clotting factors named I through XIII, not for the order in which they are used but for the order in which they were discovered, which seemed entirely immune to mnemonics and other memorization techniques.

To memorize and then truly understand these kinds of elaborate processes was a nearly insurmountable task for me. While down in Brooklyn my brother was beginning a path that would lead him to be a cardiologist in the very same office as our father, my struggles to drink from the fire hose of medical knowledge were part of what guided me to psychiatry, a field where memorizing protocols seemed less important than learning patients' stories. I imagined that grounding my experience in the human element of medicine would be the only way I could thrive.

I wondered if my residency classmates, coming from more prestigious medical schools, had shared any of my challenges. I shivered at the thought of how brilliant these people must be — and therefore, I assumed, how self-centered and entitled. Regarding the latter, I couldn't have been more wrong.

• • •

I met most of my fourteen classmates on a Sunday before orientation technically began. I spotted the group gathering about fifty feet away at an outdoor picnic spot near the entrance to the hospital. My heart skipped a beat as I began a gallop toward the group only to be stopped short by a disheveled man in a baseball cap who had stepped in front of me.

I pulled up abruptly just in time to avoid crashing into him.

"I'm sorry to bother you, sir," he said. "It's just that I'm here for—"

"Oh, that's okay. I'm just going to meet—"

He didn't let me finish or move past him.

"You see, sir, I'm here receiving my cancer treatments."

He lifted his cap to reveal a head with several distinct bald patches.

"I'm sorry to hear that."

"Well, the thing is—"

I could feel eyes beginning to look over at me from the group gathering just a stone's throw away.

"I'm sorry. I really have to—"

"Sir," he said, raising his voice somewhat. "It's just that I didn't plan ahead. The chemo messes with my memory, and I'm afraid I don't have enough gas to get home."

"Oh," was all I could think to reply.

"Could you"—he looked down at his shoes, seeming ashamed of what he was about to say next—"could you lend me ten bucks so I can fill up and get home? I'll pay you back if you just give me your address. I'll mail it back to you. I'm good for it."

"Ten dollars?" I asked.

I had the sense that I was being scammed, but I felt the

weight of my new classmates' eyes on me, and I pulled out my wallet.

"All I have is a twenty," I said remorsefully.

"That's fine. I'll take that. Just give me your address, and I'll—"

I began to feel my cheeks becoming red. I had to get out of the situation, preferably without getting into an altercation with a stranger right in front of the residents I'd be spending the next four years with.

"Take it," I said.

I practically shoved the bill into his hand in one motion and swept past him toward my group. As I gathered speed away from him, I hoped the interaction hadn't given my classmates a strange impression of me. Was giving money to strangers in need something people at Harvard did? I felt enormously self-conscious and just hoped no one had been paying attention.

"God bless you, Doctah!" the man shouted at the top of his lungs.

Everyone turned to greet me after that. The senior resident who was in charge of the day's program, Rebecca, came over and shook my hand.

"I see you met Slippery Nick," she said with a smile. "He stays in one of the group homes nearby. He doesn't have cancer, just so you know. It's a benign autoimmune condition that makes his head look like that. Did you give him money?"

"Yeah. Twenty bucks."

"Twenty! Well, don't sweat it. Once he knows that you work here and are onto the con, he stops pretty quickly. After a while, it's actually kind of nice to see him after a tough call shift or before the start of a long day."

Rebecca introduced me to the group, and one after another greeted me with a warm smile and questions about my interests in psychiatry and beyond. I explained that I had wanted to be a doctor for a very long time but that it wasn't until rotating in psychiatry as a medical student that I knew it was the best fit for me. Every patient had a story, and it seemed like the ideal treatment could almost always be determined by trying to better understand him at his core. All medical fields involve the treatment of human beings, but in psychiatry, basic humanity — rather than, say, the kidney or the skeletal structure — is the very foundation. I worried that I sounded like a hippie or an idiot, but quickly I learned that many of my new classmates felt exactly the same way. It seemed too good to be true.

The group spent a crisp Boston afternoon getting to know our newly adopted city through a carefully choreographed scavenger hunt led by Rebecca. There were items like "spotting the giant Citgo sign outside of Fenway Park" and "sit on a *Make Way for Ducklings* statue in the Public Garden." It felt silly for a bunch of self-serious and dogmatically educated young professionals to be engaging in a scavenger hunt better suited for children, but it brought us together for an entire afternoon and introduced us to the city we would call home for at least the next four years.

Rebecca's clothing and positivity reminded me of my elementary school art teacher. Again it made me question what kind of preconceived notions I had about who became Harvard psychiatrists. Throughout the journey, she made time to ask about each of us and really listened to our answers — already showing off the empathic listening skills I'd later come to recognize as fundamental tools in psychiatry.

"So where are you coming from, Adam?"

"Oh, SUNY Upstate. It's in Syracuse."

The half second of silence that followed triggered me to fill in the void with preemptive defensiveness.

"I don't know how I matched here, either."

"Well, they only match people they really want, so just keep that in mind along the way. No matter what happens, you are not here by mistake. You belong here," she said.

For the moment I actually thought maybe she was right.

As we checked off the last items on our scavenger hunt, I realized we had spent much of the day getting to know one another. Miranda had actually grown up on Long Island, like me, until her family moved to Massachusetts in her preteen years. She romanticized life on the South Shore of Long Island in the same way I thought about summer camp as an eight-year-old. Just from listening to her speak, I could tell that Erin was very bright but possessed a nervous self-doubt that was evident in every word she spoke. Maybe they *were* smarter than I was, but they were also kinder and more intellectually curious. Each person I spoke to had a charming backstory, and I couldn't wait to learn about every one of them. I looked at the group with bright, admiring eyes. Maybe this would all work out.

The next day, as we began formal orientation at the medical center, the last member of our group arrived. She had been at a friend's wedding the day before and had missed the scavenger hunt. I thought she was exquisitely beautiful and watched her move with a quiet elegance I had never seen before. I approached her and extended my hand with a grin.

"I'm Adam. Good to meet you."

She looked at my hand for what felt like an eternity before shaking it unenthusiastically and introducing herself with as few words as possible.

"Rachel," she said flatly before turning away.

Rejected and deflated, I turned away as the blood began to pool in my cheeks. All she had done was offer a reserved greeting, but it was a blow that had taken the wind out of me.

2

The Golden Class

We began residency with a weeklong orientation boot camp. On that first Monday morning, all fifteen of us showed up early and in our best attire, eager to prove ourselves worthy of our selection.

Everyone seemed stiff with anxiety. The only one who seemed calm was a woman named Gwen, who had matched into the Harvard residency straight from Harvard Medical School. No doubt it was a more natural transition for her than for the rest of us. She was even already familiar with the hospitals we would rotate through during our time at Harvard Longwood.

"Are we in, like, the basement?" Dana asked.

"We just walked in together from the street. We're at street level," Ben replied.

"No, I know, but doesn't this room feel kind of like it belongs in a basement?"

Dana and Ben knew each other from medical school and shared a vibe of familiarity.

"Gwen," Dana continued. "You were here for med school. Are all the rooms like this?"

"Like what?" she asked with a smirk.

Dana paused to contemplate her phrasing.

"Like a Realtor would describe them as *charming*."

"Well, you know, academic medical centers often do put psychiatry in the not-so-desirable locations," she replied.

"Why?"

"We make very little money for the hospital and have almost no fancy equipment that requires renovated space," I said.

I didn't know if that was actually accurate, but I was trying to sound smart. With fluorescent lights overhead and a constant buzz emanating from the old HVAC, my surroundings were not as sleek as I had expected, but somehow they seemed even more intimidating.

Miranda tried to open the conference room door again. "Yep, still locked," she said, joking. "Just thought I'd check if I had the magic touch."

Miranda was coming from another Ivy League school and bubbled with an effervescence that straddled the line between anxiety and excitement.

As we gathered around, I stole a glance at Rachel. I had been nervous that she would notice my peeking over at her, but I didn't seem to exist in her world.

We waited outside the conference room for another ten minutes before any of us realized something was amiss. After more polite but awkward small talk, Ben finally asked if we were in the right place. Dana thought to contact the program's administrative lead, Tina, and we learned that the first faculty member on the schedule had forgotten about us.

Tina rushed over from the other side of the medical center

and made note of the hectic nature of orientation week. The faster we became comfortable rolling with the punches, she said, the better off we'd be.

Tina was a bit older than us — probably in her mid-thirties — and hard not to find fascinating. She wore mostly dark clothing that leaned toward a professional goth style, and one side of her head was shaved close, while the other featured long black hair over a model's high cheekbones. Later I would learn that running the residency program was Tina's day job, while in off-hours she poured her passion into an underground Boston art scene that involved screenwriting and directing horror flicks that won awards on the local short-film circuit. For us, though, she basically ran everything. If any of us needed something, we called her. She seemed to be the only person in the entire department who knew how to get everything done. Tina was a master of *doing*.

With a few well-worded requests, Tina was able to get us started and led us into the conference room. Dr. Carol Redding, the program director, eventually came, and we all sat up straight.

"First, you all belong here. Make no mistake. We wanted you, or we wouldn't have ranked you on our Match list. Second, I know you don't feel like psychiatrists yet. That's okay. That's how it's supposed to be. In fact, that's what you're here to learn."

She commanded the room by speaking with a soft firmness that earned respect.

"Third, please don't make me ever have to call you into my office to tell you to dress more appropriately. I hate that." She paused. "And don't make me tell you what I mean by appropriate. You're grown-ups. You should know. And while we're on the topic of things you should know, I guess it's only right to tell you how excited we are that you're here. You are by far the most highly ranked class in the history of the program, and I am delighted to welcome you."

That's when Miranda whispered to me our new class identity: "*The Golden Class.*"

The phrase, offered in jest, made me feel like even more of an impostor, despite what Dr. Redding had just said. I couldn't fathom how to avoid being the anchor that weighed down and held back this magnificent group.

As Dr. Redding finished up her opening spiel, Tina began handing out our ID badges, which I saw listed me as belonging to the "Department of Physchiatry"—some sort of typographical love child between psychiatry and physiatry, I guessed.

"Physchiatry?" I said to no one.

"Maybe it's a brand-new department created just for you," Dana said, smiling.

"I guess I'll be the chair, the faculty, an administrative person, and the custodian," I replied.

When issues would arise over the next four years, I would always blame problems on internal political strife within the physchiatry department.

Next, we were introduced to Nina, the course director for a seminar called Becoming a Psychiatrist. Nina insisted that we refer to her by her first name, though our default was to call attending psychiatrists by the more formal Dr. So-and-So. She tried to explain what the course was all about, but the words—each knowable on its own—became less clear when strung together.

"It's a didactic without a formal curriculum, but we have definite goals for individual and group-level development," she explained. "We'll be using our experiences as a kind of starting-off point for how to get from where you are now to where you will be in four years. A lot of the content will involve processing our emotions, but it's definitely not a group therapy session."

"So it's a class about our feelings? *Feelings class?*" Rachel asked.

"Yeah, that works," Nina said, laughing right along with us.

I knew right away that I liked Nina, and imagined we'd be needing her in the years ahead. There were going to be a lot of complicated feelings, and I had a sense that Nina knew the way through them.

With the informal name of the seminar settled, we began to discuss some of those early feelings, and it turned out that we were all terrified.

"Does anyone else feel like they know how to be a doctor? Because I definitely don't," said Erin. "What if I kill someone?"

"I just hope there's actually someone there to catch us when we mess up, because now there are actually lives at stake when we make decisions. Why did I decide to be a doctor again?" I asked.

"What if I order two serotonergic medications without realizing it and cause serotonin syndrome? I don't know what hypertonic reflexes even look like, I mean except what I've seen on YouTube. How will I know not to do that if I've never not done it before?" asked Miranda.

The temperature in the room was rising.

"Where Ben and I went to school, I feel like we were coddled. We were little children who weren't given any hands-on responsibilities," Dana said. "Now, apparently we get thrown in and are asked to do everything as though we've been doing it all along. We're supposed to just march in there on Monday and be the doctor? I can't believe this is how it all works."

It amazed me that my classmates from Ivy League schools felt just as petrified as I did. Only Rachel seemed to take things in stride.

"It'll be fine. Just do what seems right, and if you don't know something, look it up or ask someone more senior. Medicine is pretty easy when you approach it like that," she said.

"I like that," Nina said. "One of the very nice mantras around here, coined by one of our senior faculty, is to *never worry alone.*"

And with that, for the first time all day, all of us relaxed just a little bit.

3

It Will Pass, and You Will Be Fine

From four feet away, she looked dead. What would I do if she actually was? It was my first clinical day on the neurology consult team, and I had no idea what I was doing. My newly ironed white coat felt like it weighed a ton, and my palms were sweating nonstop. I wondered if it would be the day that the entire fantastical Harvard endeavor would come crashing down due to my obviously lacking medical skills.

Each of us began the first year of training on a different specialty service. I was assigned to start the very first Monday after orientation on the neurology service — a two-month requirement for all first-year psychiatry residents. On that first morning, I was told by the new senior resident to assess an elderly woman who had experienced a period of *asystole* — her heart had stopped for several minutes before being shocked back to some version of life. Neurology had been consulted specifically to determine if this woman was, in fact, brain-dead.

I had no real idea how to complete the task that had been assigned to me, but I was eager and resourceful enough to know that hospitals generally have access to very thorough educational subscription services that could direct me in how to tell if someone was brain-dead. I logged on to a database called UpToDate and learned the various maneuvers I would have to perform to determine if this woman had any ongoing brain activity. I was entirely unsupervised for this pre-rounds period and spent the better part of an hour trying to decipher the steps laid out in the UpToDate protocol.

I inched closer to the body, perfectly still on the ICU bed. The intensity of the fluorescent lights made me want to shield my eyes. No one else was nearby, though the chimes and dings from the surrounding machines offered me a kind of audience.

The woman was being pulsed with cold fluids in her veins to keep her temperature down. Her skin was ice cold.

Upon slightly closer inspection, she still seemed dead to me. As I was about to begin my exam, the primary medical team arrived with a flourish.

"Oh, you're from Neurology. Great! Thanks for coming," the team leader said.

"Oh, no, I'm just the intern," I replied sheepishly.

"Just the intern? Today's your first day? Say it with pride, my man! You're the doctor now! We'll come back in twenty minutes. Drop a note in the chart with your assessment."

The group scurried off to the next patient on their rounds, and I returned to my examination. I leaned in closer to the patient and felt her cold breath on my arm as I tried to take the carotid pulse in her neck. It was stronger than I'd anticipated. I put my fingers into her open hands, and she instinctively gripped them. When I lifted her eyelids and turned her head from side to side, her eyes opposed the motion. This was a good sign according to UpToDate.

To my own amazement, I introduced myself aloud as "Dr.

Stern" for the very first time. I apologized about needing to shine some light in her eyes and almost jumped into the air when I found normal pupillary responses. The old woman then withdrew from painful stimuli when I pricked her fingertip, and I apologized again. I had my answer; it seemed that neither of us was brain-dead after all.

After I completed the not-dead woman's consult note, I went on to the next patient, a woman who'd been brought to the hospital with something called transient global amnesia, a rare but usually harmless disorder where the brain spontaneously and temporarily loses the ability to form new memories. I had what I thought was a completely engaging conversation with the patient about her upbringing and her marriage and the job she hated, and then I stepped out of the room to grab a reflex hammer. When I came back in, it was like I had never been there. She reintroduced herself and started to explain that she could tell she was in a hospital but couldn't entirely tell why she was there. Her husband motioned to a sheet of paper that read:

You are in the hospital because you are having a problem with your memory.

It will pass, and you will be fine.

I found the whole interaction utterly fascinating, like a scene out of *Memento* but without the weight of murder and tragedy. Over the coming hours, her brain healed itself without any help from me. As with a resolving seizure or a migraine that finally goes away, this woman's brain was able to reset itself. I was relieved to see that my uselessness wasn't a detriment to her outcome.

The rest of the day was a blur as I ran on the adrenaline of being a *real-life doctor*. The impostor syndrome all early-career docs experience was strong that day, but it was overpowered by my pride and eagerness in every action I took.

I did three more consults, for a suspected stroke, a patient who had been hit in the head by a falling branch, and a patient with worsening ALS who was struggling to breathe on his own. Somehow, I felt like I found a way to connect with each patient, though there was one low point. The patient with ALS had in a tracheostomy tube to help him breathe. It wasn't properly capped, and during my examination he coughed up some secretions that sailed across the room and against all odds landed on my mouth and chin. Panicked, I excused myself and rinsed the area thoroughly before asking the nurses for mouthwash and soap. Convinced that I had just been exposed to HIV, hepatitis B, and Ebola all at once, I told one of the nurses what had happened, thinking she would surely advise me to go to Occupational Health for a kind of exposure prophylaxis work-up, but she just shrugged.

"Occupational hazard, dear."

I probably should have gone anyway, but there was no actual evidence of infectious disease exposure, and it felt like there was too much left to do. Going to Occupational Health now seemed like a grade schooler going to the school nurse with a bellyache, so I chose to soldier on with the day's work instead.

At 6 p.m. I changed from my work clothes into scrubs and headed to the locked psychiatry unit on 4 South. Even though I was rotating through Neurology, I would still be on overnight call in Psychiatry. Because it was my first night on duty, a teaching-call resident was tasked with showing me the way and making sure I had someone to answer my questions. I was relieved it turned out to be the senior resident, Rebecca, whom I had already met.

When I saw her outside the unit as scheduled, she gave me a hug, which took me by surprise.

"You'll need that," she said before launching into an extensive description of the machinations of an inpatient psychiatry unit.

"Look, when you're on call overnight, every single person around you is someone who is potentially looking to take advantage of you. They can all hurt you. You need to have armor on. I like to imagine that I'm wearing an invisible hat that has middle fingers all the way around."

"Wow. Okay. You sound sort of burnt out," I replied.

"You have a keen clinical eye, Dr. Stern," she said, placing a hand on my shoulder.

We entered the unit with the swipe of a badge. To the uninitiated, it can be an overwhelming experience to walk onto a locked psych unit. We were greeted with a vacant stare from the checks person, whose job was to lay eyes on each patient every fifteen minutes to ensure everyone was accounted for and in a reasonably safe state. There were lots of functions to the inpatient unit including therapeutic care, group therapy, occupational and physical therapy, and, of course, medication optimization. But the very first purpose of the inpatient unit was to keep the acutely unsafe from harming themselves or anyone else. The checks person acted as our eyes and ears in this endeavor.

Next, we walked by a middle-aged woman talking to herself who did not make eye contact.

"It's not because of that. It's the other thing," the woman said. "No! It's the *other* thing!" she then shouted to herself.

"That's Ginger. She's been in and out of here for twenty years, but no one's quite figured out what the other thing is."

We entered the nurses' station, where there were rows of computers and a window into the dayroom, where patients hung out when there weren't clinical activities going on. I gazed in and saw an assortment of visibly suffering individuals. There was a young woman with anorexia nervosa, emaciated to the bones and still studying intently at the small table off to the left. To her right, I took note of a man I guessed to be in his fifties with a big bushy white beard, who was exhibiting what is known as *waxy flexibility,* a characteristic of patients with severe catatonia that leads them to become frozen in unusual positions.

Rebecca noticed me staring.

"Roger's actually doing a lot better than when he came in. The ECT is helping—shock therapy. Sometimes he'll be relatively loose and agile for hours at a time now. If he gets really bad, IM lorazepam, two milligrams."

"IM?"

She motioned to her buttocks with an imaginary needle and mouthed the words *intramuscular.*

She shifted her gaze to the far corner of the room.

"Those two, I'm worried about."

She motioned to two relatively benign-looking individuals holding hands.

"Why?"

"They have a Romeo and Juliet vibe going on. Met on the

unit, inseparable. On several occasions they have been found by the checks person in the shower together. Keep an eye on them."

She began to pack up her belongings.

"I think you've got it from here. Page me if anything comes up."

"Wait, no, you can't go. What do I do if we find them in the shower together? What do I do if—"

"You'll figure it out!" she shouted as she made her way to the unit door. "And if you can't, just page the senior resident."

"What if they don't respond?"

"You're the doctor now!"

"Never worry alone," I muttered to myself, wondering how I'd manage through the night without anything catastrophic happening on my watch.

I had successfully completed three new admissions to the unit when I started to feel like I'd found my groove. The admission process, it turned out, was mostly paperwork. There was the safety-plan form, where we documented what to do if the patient described feeling unsafe, the medication-reconciliation form to confirm we were giving the right medications, and, perhaps most important, the conditional-voluntary form. This document allowed the patient to sign in voluntarily. Patients who refused would still be kept on the unit against their will for up to three business days or until a judge ruled on the case. This meant it could be more than five days if the patient came onto the unit on a Friday evening.

I realized when admitting the patients that they had all been spending time—often extensive time—in the emergency department and mainly seemed relieved to finally settle into a room. Admission was not the time to talk about their feelings or childhood or whatever else seemed like the *psychiatric* thing to do. Even against the monumental forces of suicidality and psychosis, the power of physical and emotional exhaustion was immense. In all three admissions, the patients just wanted to get through the interview as quickly as possible and go to bed.

I had finished the last bit of paperwork and decided to try to get some sleep in the overnight call room, a small, bare-bones space with a twin bed, computer, and phone, surrounded by painted concrete walls and tiled floors. Just as I turned out the light, the first of a torrent of pages came in.

Roger needed lorazepam. One of the new admissions was missing her order for quetiapine, an antipsychotic. Several patients couldn't sleep and were requesting a variety of drugs. But all of those issues would have to wait because Ginger was no longer self-dialoguing at a reasonable volume. It was 3 a.m., and she was shouting so loudly I could hear her on the other end of the line as the nurse matter-of-factly asked me to return to the unit to see what I could do.

When I arrived, I saw her in the dayroom irritably pacing, while poor Roger sat silently at his center table in a completely unnatural position, his arm crooked above and behind his ear, seemingly unaware of the chaos all around him.

"Son of a bitch! The other one! It's the other one! You motherfucking motherfucker. The other one. For Devil's Christ."

I checked Ginger's pre-written sign-out — a detailed instruction set for each patient from the day team about how to handle agitation or other likely problems. She had already been given all of her PRNs — *pro re nata* medications, which loosely translates from Latin to "as needed." I would have to figure out a way to defuse the situation or take the undesirable step of ordering meds above what had been advised by the day team.

"Ginger," I said quietly as I entered the room. "I'm Dr. Stern. I am the psychiatrist on call tonight."

She stopped in her tracks and then began approaching me quickly. It looked like she might pounce on me, and I instinctively tensed up. She stopped an appropriate three feet from me. From this distance, I realized that she wasn't aggressive. She was afraid.

"Can you help me?"

"I hope so. Can we go to the Sensory Room and talk about what's going on?"

I led her down the hall to a room that was designed to be calming, with arboreal wallpaper.

I motioned for her to have a seat on the couch farthest into the room and took a chair closer to the exit. For self-protection, I had learned during orientation to always be closer to the exit than a patient, but under certain circumstances it could be the worst placement because it can make a paranoid patient feel trapped. Ginger didn't strike me as delusional in that moment though, only distressed.

"It's very late, Ginger. What's keeping you awake?"

"It's the other one. The other one," she repeated, this time in a hushed tone as though she were sharing secret information.

"The other one?"

"The other one."

"Tell me more about the problem."

Ginger went on to speak in a trajectory I could not follow. She spoke of her childhood home and her parents and a teacher named Mrs. Vanderpost. She motioned to both sides of her body and spoke of her medications and the other doctors taking care of her. I couldn't seem to break into the conversation; it was all happening so quickly. Then she stood up with a start, and again my entire body tensed, ready for an attack that didn't come. She walked slowly to a mirror on the wall. She motioned to the hospital ID bracelet on her left hand and pointed to it in her reflection. She motioned to an emblem on her gown and the same one in her reflection. She lowered her head and showed me the part on the right side of her hair and again pointed to the mirror.

"Demons," she said despondently. "They always put it on the other one. They always take something good and make it bad." I wasn't sure that I was on the right track, but I suspected that wrapped up in all of the psychotic thought processes that plagued her, a part of Ginger was concretely noticing that everything in her reflection was being displayed back to her on the reverse side, and her illness was in some way leading her to conclude it was the work of demons.

I said the only thing that came to my mind.

"You're safe here, Ginger."

I knew I had no power over demons, but I did have an idea. It came from a deep corner of my mind, from a memory of when I was twelve. I had always parted my hair on the left side, and saw someone at the hairdresser getting a trim and parting it on the right side. I commented to my mother that I wondered how I would look with it on that side, and she said that I would look exactly how I see myself in the mirror, but that people looked much better with parts on the left. While the last comment swept past me without a thought, I felt shocked to realize that my entire life I had been looking at myself backwards in the mirror and in fact in the real world I was walking around looking like myself in reverse. I wondered if Ginger had been experiencing something similar but more intense for decades.

"Stay here for a moment," I said, walking purposefully out of the room. Once in the hall, I ran to the nurses' station and with great urgency asked if anyone had a pocket mirror. A nurse fished one out of her purse, and I took it back to Ginger.

She was still positioned in front of the mirror on the wall. I opened the pocket mirror behind her head.

"Look at your reflection through this," I said.

As her eyes tracked up at the mirror in front of her and then back from the reflection I was showing her, the anguish in her face dissipated.

"You're safe here," I repeated.

A sheepish smile crept onto her face.

"Safe," she said.

I walked Ginger back to her room, where she got into bed without another word.

When I got back to the dayroom, a nurse had already administered Roger's lorazepam. He was sitting at his table, eating a banana and drinking orange juice with the paper in front of him like it was a lazy Sunday morning. He was no longer stiff.

"Hey, Doc. Great juice in this place!"

"Happy you like it, Roger."

I walked out of the dayroom and the unit and peered out the single window I could see. The sun was just beginning to rise.

David: hey

me: hey

me: slow day here, What's new with my brother?

David: same here

me: i'm working 8AM tomorrow-8AM Sun, and there are a whole lot of beds open on the unit which means it's fairly likely there will be a whole lot of admissions for me to do.

David: good times

me: neuro consult service next week and goes for the next 2 months. overnight psych call Wed.

David: i HATE having thank you notes to write

David: they loom over me like almost nothing else

me: Probably should have thought of that before getting married

me: yo

David: hey

David: i'm on night float

me: i'm on call as well, what's the nightfloat schedule like?

David: i have one more week

David: it's 8pm-7am(ish) six days a week

David: shhhhhh . . . but so far, 0 admissions 0 codes

I bumped into one of Dad's partners from the office.

he quizzed me on an ekg and then talked to me about pacemakers

he loves teaching

David: it's nice

me: i'm going to quiz him on atypical antipsychotics next time I see him

David: indeed

me: i'm on 24 hr call tomorrow but the unit should be full (there are 2 beds open right now) so it'll all depend how busy the ED is

me: I wish there was a TV in this call room

me: there's a Mets game on ESPN right now

David: i have a TV . . . but no espn

David: it's torture

me: i could watch in the TV room on the psych unit, but things might get hairy

David: i'm tempted to pull a HOUSE MD and watch in a patient's room . . . but i'm worried about a) how it'd look, and b) whether they get espn at all

David: this Mets team is just a huge huge HUGE disappointment all season long

David: what can you tell me about hypomania?

me: what do you mean

David: i was accused of being hypomanic — i want a clever response

me: i don't know about clever . . . hypomania's interesting from an academic standpoint. productive, requires little sleep, happy bordering on too happy. Mania is the nightmare that it almost always becomes.

David: well there's not much I can do with that

David: i want the Mets DOMINATING the rest of the season

David: getting their averages back above .300 and such

me: not hopeful about that.

me: uh oh. Pager. Talk to you later.

4

First Paycheck Party

Residents' status in the world is an inherent contradiction. As people who gave away their mid-twenties to medical school, we are burdened with the unreasonable weight of being responsible for life and death while also being totally developmentally stunted. Most of my cohort had never had a full-time job before starting residency, so at the end of our first two weeks, we were receiving the first significant paychecks of our lives. It was an event that inspired so much pride that the residency had an informal tradition of sponsoring a "first paycheck party" at a nearby Mexican restaurant.

The fifteen of us gathered in celebration. We were a diverse bunch — married and single, city folk and country, from all backgrounds. There were twice as many women as men, which wasn't uncommon in psychiatry classes. We didn't feel quite at home with one another yet, but the first paycheck party was aimed at correcting that. By the end of our four years together, we'd be family, but until that night it had mostly felt like we'd been alerted by an ancestry service that we were all distant cousins — *Congratulations on being stuck with these strangers! Hope you like them!*

When I showed up at the taqueria, there were only a handful of people gathered around small, standing tables. I arrived just as Miranda was coming back from the bar with two drinks.

"Hey!" I said, probably too enthusiastically. "You here with someone?"

I motioned to the two drinks in front of her.

"Oh, hi! No, these are just for me."

I liked Miranda right away. She exuded an earthy friendliness paired with obvious intelligence and inquisitiveness, and her Long Island accent made me feel at home.

"I have a feeling we're going to end up spending down most of our paycheck right at this bar," I said, coming back from the bar with nine fewer dollars in my pocket.

"There are worse ways for it to go," she replied.

Rachel had now arrived and joined our group.

"I hate things like this," she said. "Small talk is the worst."

"Oh, I love it," Miranda said.

It was the first data point in a series involving Miranda and Rachel that revealed their yin and yang personalities; it seemed like they couldn't possibly get along, yet somehow they worked perfectly together.

"Okay, listen, you social butterflies," Rachel said sternly, "let's make a pact. I don't want to be left alone with any creepers, so at any given time I want one of you to be right here."

I nodded in agreement, flattered that I wasn't a creeper in Rachel's eyes.

"And rule number two," she continued. "None of us is going to make out here tonight."

Miranda and I nodded obediently, though now that Rachel had placed the idea in my head, my mind began to imagine the possibilities. Was she talking about me and Miranda or me and her or the two of them?

By our third round of drinks, we were socially lubricated. Our trio had ventured away from our initial table to chat for some time with Dana and Ben and a new arrival, Erin. She had arrived with her impossibly good-looking husband, Bobby, but he left after about five minutes.

I noticed several of the women in attendance ogling him as he exited. I turned my smoldering jealousy into polite conversation.

"So, what's Bobby's deal?" I asked as Rachel and Miranda listened intently.

"We met in high school," Erin replied. "Got married just after college. His family was very supportive during graduate school — he got his PhD in applied mathematics just last year — but now that we're here, I think he misses them a lot. I can never seem to get him to socialize for more than a few minutes."

She told us that they had decided that Erin's residency match would determine where they lived for the next four years, but then their next move would be for him and his career. It was a reasonable bargain for these two highly specialized academics, but it did mean that Bobby would be stuck in a city he had no love for trying to find work below his degree qualifications just to give him something to do.

"He already seems miserable, and we've just been here for a few weeks. I don't know how we're going to make it through four years of this."

I felt uneasy sitting with such an honest disclosure of her husband's unhappiness. Later in my training I would know how to sit with the tension of misery, but I didn't yet. Miranda stepped in to attempt a redirection.

"So I had to call an interesting consult last night," she said out of nowhere.

I pleaded with my eyes for her to keep going.

"There's a woman on the unit admitted with paranoid schizophrenia."

"Do you think we should be talking about this out in public?" Erin asked.

I was beginning to sense that Erin might be more straight-laced than the rest of us.

"Sure, it's just peer supervision," Rachel said with a smile, prompting Miranda to continue.

"So I get this page from nursing — *so-and-so needs you to examine her privately*. And I'm just like okay, whatever, let's do it. So I take her to the exam room, and she tells me that she thinks she's growing a penis."

"Get the *New England Journal* on the horn!" I said.

"I told her that wasn't possible, but she asked if I could look anyway. Okay, I thought, I can't not look if the patient is concerned, right? I'm a doctor now. I have to look. So I examine the patient's privates, and she wasn't growing a penis, but there was definitely *something* growing there."

"Ew. This is not why I went into psychiatry," Rachel said.

"Come on," Erin said, "have compassion for the patient! It's a really interesting intersection of her delusions and physicality. So what did you do?"

"Well, I knew I was in over my head. What could it even be? Vaginal cyst? Who knows. So I paged GYN for a consult."

"What did you say in the page?" I asked.

"Rule out penis, of course. Got the job done. They came for the consult, and the growing penis was no longer mine to worry about."

"I think I would feel like just the dumbest, most useless doctor in the history of the medical center," I said. "*I* would feel that way. You shouldn't."

"Oh, I was mortified. Our department's reputation definitely took a bit of a hit that night."

"How did the gynecologist react?"

"She was really nice, actually. She kept trying to tell me about the differential diagnosis for a small vaginal mass, though, and I kept trying to politely end the conversation."

"You didn't specialize in psychiatry to focus on that?" Rachel asked jokingly.

"Sometimes a vagina is just a vagina," I said, channeling Freud.

"Do *not* quote Freud at a bar," Rachel said, warning.

"It's okay," Miranda said. "As I understand it, Freud was only obsessed with male genitalia."

There was a short pause.

"I wonder how that patient is doing," Erin finally said.

None of us could very easily leave our work at the hospital, it seemed.

"No more work talk tonight, people. Let's get another round," Rachel commanded.

The rest of the party was light and cheery. I began to think that maybe I could get through these next four years with this group by my side. As I headed off to sleep that night, I hoped Rachel's stance on "no one making out" might end up being more of a guideline than a rule.

5

The Mythical Day Team

Work on the inpatient service overnight was defined by two primary goals: keep everyone alive, and make as few changes to the day team's plan as possible. New residents working overnight were valued members of the team, but our work was almost meant to be invisible. All of us, including us interns, understood our limitations with regard to effective treatment and discharge planning. If an overnight resident kept the patient list relatively intact with no major mistakes being made, it was considered a good shift. When viewed through that lens, the job was manageable, though still exhausting. Each patient came with a written and verbally communicated sign-out from the mythical *day team*, which had the luxury of comprising a resident, a social worker, an attending, and a nurse. To the overnight resident flying solo with just a single senior resident seeing patients in the emergency department as support, the day team's manpower seemed almost comically abundant. Before becoming part of the day team, I imagined it as a kind of omniscient force of nature, where patients' most intense symptoms, such as suicidality and psychosis, would be figured out and effectively treated over the course of their three-to-seven-day admission. Throughout my nights on call, I found myself saying, "You'll have to discuss that with your day team" over and over again when patients came looking to me for answers. What I left unsaid in my response was what I implicitly knew: the day team knows what they're doing, and I do not.

When I eventually rotated onto the day team as part of my first-year schedule, it became clear that the slow, almost cerebral treatment planning and coordination of care I had imagined was not realistic. The members of the day team were just doing their best, often on the fly in a very chaotic environment. The primary goal was to get people stabilized and set up for a safe and successful return to life outside the unit. By definition, patients only truly met an inpatient level of care on the locked ward if they were imminently unsafe to themselves or others. In Massachusetts this can be established legally when the patient voluntarily admits herself, or through a Section 12 form, which is a sheet of paper that gives a psychiatric team three business days to stabilize the patient against her will before she can either be discharged (against medical advice) or be brought to court to advocate for ongoing treatment through a guardianship. I learned quickly that I, along with most psychiatrists, do not like treating patients against their will. We look for any avenue available to avoid it, but sometimes it is undeniable that patients are at imminent risk of self-harm or harm to others. When that is the case, we do what we must by going forward with involuntary treatment despite how awful it can feel. Yet patients on the unit would frequently accuse psychiatrists of having ulterior motives for keeping them hospitalized. That accusation always cut me deep because there was nothing I wanted less than to take away someone's freedoms.

On our first day as members of the day team, Erin and I (who were rotating together) were shown around by the lead attending on the unit, Dr. Song.

"Welcome, my friends," he said. "I know that it does not look like much, but 4 South will become somewhat of a home to you, and you will learn a great deal about psychiatry and humanity here."

Song had a reputation for being a bit eccentric but bighearted. He showed up to work barely in time for morning rounds, wearing bicycle shorts that left little to the imagination. When he entered a meeting, though, his eyes instantly met whoever was speaking and dug straight into their soul. He had a preternatural ability to understand people, even, and maybe especially, patients who had trouble interacting with the world around them.

"Patients are people. We are people. Be a person with your patients, and you are already halfway there. Whether we use olanzapine or risperidone for hallucinations is less important than whether we use empathic listening. Ah, and speaking of empathic listening: this is Crystal, our team's social worker. She is one of the best listeners in the galaxy, and the unit simply would not run without her."

"That's true," Crystal replied, "but I do not tire of hearing it, Song."

Crystal and Dr. Song had been working together for so long that they had a natural rhythm and hardly needed to speak at all to understand each other.

Crystal walked us over to the big board and showed us the list, which by now we were well versed in navigating from our overnight calls.

"Let's make sure you each get a good variety of patients. Hmmm . . ."

She began writing illegible notes on the board in dry-erase marker.

"Erin and Adam get these five and these five. You two decide who carries which group," she concluded. "First group meets at ten a.m. for rounds, and second group meets at eleven thirty a.m."

I deferred to Erin and ended up with a group that included several patients I already knew. There was Ginger (suffering through her mirror psychosis), Roger (catatonia), Jane (anorexia), and Paul (the Romeo of our unit couple). There was also a new admission, who came in overnight with acute mania — Deborah.

We gathered for rounds in one of the unit's conference rooms, and the first patient to come in was Paul. He looked terrible, notably worse than just a few days earlier. His hair was greasy, and his shoulders were slumped. When he sat down, I noticed several stains on his shirt. This was a dramatic deterioration.

"You're up, Dr. Stern," Song said.

"Paul," I began. "How are you today?"

He sat in silence, staring at the tiled floor. I waited for three seconds and looked to Song for help. He held up seven fingers — instructions to wait seven seconds before following up on a question. It felt like an eternity.

"Paul. I'm concerned that —"

"She's gone," he said flatly. "Do you know what it feels like to be numb your entire life and then to wake up?"

I shook my head.

"It feels like birth, and this feels like death."

"Your girlfriend left?"

He nodded.

"The unit?"

"The country. She went back to Sweden with her stepfather. My life has no purpose without her."

My eyes looked to Song for guidance. I hadn't learned a single lesson in medical school about lost love.

After we sat in silence for another seven seconds and I still had no idea what to say, Song mercifully put me out of my misery.

"Love is a powerful force, Paul. You feel tremendous loss. That is understandable."

Paul began to nod and even looked up at Song.

"You are still having suicidal thoughts?"

He nodded.

"Then you will remain here with us, and this is what I want you to do. Every day, you will awake at seven a.m. and spend an hour in the dayroom. Talk to every person that comes in, even if you do not feel like it — especially if you do not feel like it. Talk to them and have a full breakfast. By late morning, you will have participated in at least one activity group with the occupational therapists. We will check on your progress. Understand?"

He nodded, thanked Song, and left the room.

I stared at him in disbelief until he disappeared into the hall.

"Breakfast and an occupational group each day?" I asked.

"You think an antidepressant would work for him, Dr. Stern?"

"Well, no."

"Nor do I. This young man has a fractured self. It is so incomplete that he requires another to feel whole. When it is taken away, he tends to attempt suicide as he did before being admitted here. He attaches to others as a way to feel complete. When those bonds rupture, as they inevitably do, he cracks even more."

"So how can we help him?"

"Time, Dr. Stern. If we give the young man the gift of time on the unit and hold his despair with him, it allows his self to re-form and for other, potentially stronger bonds to take hold. We will not cure this man's crisis of identity, but we can hold him up until he finds his footing."

I didn't have time to be blown away. Roger was walking in the door already. Seeing him as a fully functioning, emotive, even charming guy after having seen his waxy flexibility on my first night on call was remarkable. I couldn't believe how quickly, with the right medication choice, a patient could transform from a rigid statue-like figure to a fully functioning person. After a brief check-in, we all felt like he had made enough progress to continue as an outpatient. The same could be said of Ginger, who had made good strides since our first night together, with less self-dialogue and no periods of severe agitation. At the end of our meeting, she leaned in and showed me a tiny pocket mirror she kept with her.

"Thank you," she whispered.

Then came Jane. At twenty-one years of age and with an intellect and work ethic that had gotten her into Harvard, Jane was devastated by anorexia. She had only made it one semester before her low body weight drew suspicions among her faculty advisors. She engaged in outpatient care for a year and parried well-meaning threats from her parents, but still she didn't eat enough to sustain a healthy weight. Little by little, she was wasting away. At the time of her admission six weeks earlier, her body had stopped menstruating and soft blond hair had begun to grow on her cheeks. She was on an eating disorders protocol, which meant that her calories were being strictly counted and, though she was weighed every day, she was not allowed to see or know her weight. A judge had ordered her for ongoing mandatory treatment, though on the unit she continued to refuse to eat. At seventy-eight pounds, Jane was dangerously close to needing a feeding tube, a gruesome experience and particularly cruel for someone who can't bring herself to eat.

I hated the idea of working against the patient — even in her own best long-term interest — by ordering a feeding tube when she didn't want one. Treating anyone against their will seemed counter to the image I had of what it meant to be a doctor — the character I had conjured up from imagining my father's daily work in cardiology. Patients come with problems looking for help, and you help them. It seemed straightforward enough, but I was learning that the real-world experience of medicine, and psychiatry in particular, was quite different.

"You're the new guy?" she asked, looking at me.

"I'm the new guy," I said.

"Should we cut to the end? I can tell you how this is going to go to save us all some time."

"How is it going to go?" I asked sincerely.

"First you're going to project empathy. You need to find a way to connect with me, right? Then you'll try to find little ways to nudge me toward eating my prescribed meals, right? And then you'll find out that I'm not eating my prescribed meals and become frustrated. Then, maybe just as we finally have our first honest conversation, you'll rotate off service and a new guy or woman will come in and we'll start the whole fucking thing over again."

"Do you think there's anything I can do to break the cycle?" I asked dumbly.

"Suicide?" she said in response.

We sat in silence for the allotted seven seconds.

"I definitely do not have the answer, but I hope that together we can come up with something."

"Ding, ding, ding! We've arrived at false empathy! I'm going now. I'll see you at the family meeting tomorrow."

As she walked away, I noticed the dozens of superficial scratches on the backs of her thighs.

"The power of eating disorders makes the most seasoned psychiatrist a humble servant," Song said sadly. "Go to the family meeting tomorrow, and try to align with the patient in front of her parents."

I nodded.

"Next up is Deborah. Fifty-four-year-old woman with a history of fourteen admissions due to bipolar disorder, admitted overnight with acute mania. Buckle up, Dr. Stern."

6

Mania

The spirit of the person pacing around the room seemed to be on fire — pulsating, flickering, burning, and red hot. I had never seen anything quite like it. She couldn't sit still, even for a moment. Her body shook as her mouth tried to keep up with the overflow of her manic ideas.

"This hospital is named for me. This is the Deborah Debingson Deborahville Hospital for the Deborahs. I invented hospitals. You are not doctors. You are doctors of Deborahs. Sexy. Sex. You."

She made direct eye contact with me.

"Can you please have a seat, Deborah?"

"Seats are cushions for the soul. Souls lead us to salvation. I want to be salvaged. Ralvaged. Ravaged. Will you?"

"She was found like this outside of the coffee shop on Longwood propositioning passersby," Crystal reported, looking at the ER note. "Hasn't slept in days per ex-husband's report. When she's not like this, she's an attorney downtown. Tax law."

"I could use a good tax lawyer," Song commented. "Well, what do you want to do, Dr. Stern?"

"Deborah, won't you sit and talk with us for a bit?"

She continued to pace furiously around the room.

"We'd like to start some medicine that will help you rest—"

But she had found the door, and her momentum, previously bouncing her around the room as though in a pinball machine, had propelled her out into the hall.

"Sleep is the answer, Dr. Stern," Song said. "The antidote to mania is simply sleep, but she cannot. What do you want to do?"

"Give her quetiapine."

"And?"

"Lorazepam?"

"Yes, and what for the longer term?"

"Lithium, maybe?"

I knew lithium was a staple in mood stabilization for patients with bipolar disorder, but this patient's presentation made me feel so far in over my head that I was completely unsure of myself.

"Lithium maybe? There is no such thing. There is only lithium. Make it so, Dr. Stern."

Just as we gathered to exit the room, the door creaked open about eight inches. Deborah's head poked in.

"You are very good-looking," she said to me before ducking back out into the hall.

Dr. Song turned his gaze to me.

"You are very good-looking?" he asked.

"I'm as surprised to learn this as you are, Dr. Song."

"Don't get a big head about it. Erotomania is very common in bipolar disorder. Perhaps we can use these kinds of declarations as a marker for her improvement."

"Am I supposed to ask her to rate my attractiveness each day?" I asked.

"We try to avoid lawsuits here, Dr. Stern. She will let us know with her spontaneous behavior."

I met Erin for lunch.

"It seems like just when I think I'm starting to get the hang of it, the floor goes out from under me," she said.

"What happened?"

"Nothing really. The last patient of the day just asked me why I thought she should go on living, and I froze. Ask me about the pharmacokinetics of chlordiazepoxide and I'm in my comfort zone, but ask me about the meaning of life and I'm a deer in headlights."

"I'm sure it wasn't that bad."

"The patient actually said it was. She said, 'You don't know either. It's that bad.'"

"The part of this job that's really taken me by surprise," I replied, "is that it really matters what words we use, and more important, how we use them. Like, this manic woman told me that I'm so good-looking. Am I supposed to say thank you? Am I *not* supposed to say thank you?"

"You're probably supposed to explore with the patient why she's crossing that professional boundary with you," Erin replied.

"Easy for you to say. She was bouncing off the walls, and I couldn't keep her in the room for more than two minutes."

"I had the exact opposite experience with one of my patients. She wouldn't get out of bed. She said she just couldn't do it emotionally—well, like her emotional distress was making it impossible for her body to get up and walk with me to a room for the interview—and I had no idea what to do either. My instinct was to sit at the foot of her bed like you might with a child who wasn't feeling well, but I knew that was wrong."

"What did you do?"

"I excused myself and asked Song about it."

"What did he say?"

"He said that it's always good to be at the patient's level."

"What does that mean? You're supposed to pull up a bed and lie down too?" I said, joking.

"I think a chair did the job. I sat down a few feet from her, and we chatted while she had her head on the pillow."

"The amount of stuff we never covered in med school is mind-blowing," I said.

My pager buzzed.

Deborah MRN 062584 requesting to speak with you.

I apologized to Erin and excused myself. I walked down the hall, up the stairs three flights, and then swiped back onto the unit. Deborah was waiting for me at the black painted line—a barrier by the door that patients are not supposed to cross, there to try to prevent elopement.

"What is it, Deborah? Are you okay?"

I could see she was still a ball of combusting energy.

"You're very good-looking. I just need you to know that you are my doctor and you are very good-looking."

"Actually, I'm glad you brought that up. I wanted to mention that because we're engaged in a professional, clinical relationship, it's not really appropriate for us to speak this way.

I appreciate your kind thoughts, but let's try to maintain our boundaries."

Her face looked as though I'd just turned her down for the prom. I felt awful as she raced back down the hallway.

About an hour later I was walking by the nurses' station when I was given a letter.

"You're in trouble," the nurse sang, handing me the folded note.

Dear Dr. Stern, we are clearly meant for one another. I understand why you must not act in accordance with your feelings in public, but please just know that I know how you really feel. One day, the world will see that we are all the same. You are not a doctor and I am not a patient. We are just people in love. Love is all we need. Sincerely, Deborah

I left the unit and headed straight for Song's office.

"What am I supposed to do with this? Please. Teach me."

"You have become the center of her erotomanic transference." He shrugged. "It will fade as the medications begin to take hold."

"And until then?"

"You go home around six, right? Just execute a bit of benign redirection until then, and I bet by morning, after some chemically assisted sleep, your shine will have dimmed, Dr. Stern." He paused. "You're not *that* good-looking."

I went back onto the unit and spent the rest of the afternoon avoiding eye contact. At 5:30 p.m. Deborah found me in the hall.

"Did you get my letter?"

"I did. It seems like you're having very strong feelings right now. What I suggest is that we—"

"But about what I said. Am I right? Do you feel like I feel? I can keep it a secret."

"Deborah, it's not appropriate," I said gently. "You're my patient. Let's just focus on your treatment and get you feeling better as soon as we ca—"

"Oh, I understand. Okay."

This time her face looked as though I had told her I'd murdered her beloved dog. She walked down the hall to her room, and I went to work on my notes in the residents' room.

I watched the clock struggle to make it to 6 p.m. Sometimes I could have sworn the minute hand was moving backwards. It read 5:58 when the overhead turned on for an announcement.

"Code blue, 4S room 23. Repeat code blue, 4S room 23. All team members please respond."

Erin and I ran out of the office and dashed down the winding corridor to the main hallway. There was already a commotion outside of room 23. As we entered, I saw Deborah's name on the doorway. I turned my head and saw that she was lying on the floor with her bedsheets wrapped around her neck. Her lips were turning blue.

"Oh God," I said as the code team burst into the room and assessed her.

The blood drained from my face as I leaned against the door watching them work.

"She has a radial pulse! She's breathing," the team leader announced as the sheets were loosened.

Deborah had been leaned up against the base of her bed and struggled to say something.

"The doctor. I want to see the doctor." She was wheezing.

The code team leader approached her.

"I'm Dr. Holzinger."

"Not you!" she sneered as her eyes searched the room, finally spotting me by the door. "Dr. Stern!"

I approached sheepishly.

"Hi, Deborah. Are you okay?"

"I am now."

"Deborah, we're going to have you stay in the Quiet Room for a bit, just so we can make sure you're okay. You frightened me, you know. I hope you won't do that again."

"Will you stay with me?" she asked with tears in her eyes.

"For a while. I'll be right outside your door until you fall asleep."

"Caring, too. Wow," she said.

I could see in her eyes the meds were just starting to take hold. By the time we reached the Quiet Room, her manic state had been transformed. She would go on to sleep for the next fifteen hours, and when she awoke, I was just her doctor again.

Managing a full caseload of patients on 4 South was like working in a firestorm. By the time one flame was extinguished, another area had been set ablaze. New to the work, we needed the space and time to decompress and process what we experienced. Our program had us work only half a day at clinical sites on Wednesdays, followed by half a day of lectures and Feelings class. That's when we all gathered in the lower-level conference room, exhausted and relieved. It was usually the first time in the course of the week that we felt like we could let our guard down. Managing patients was a never-ending maelstrom of chaos and uncertainty, but school—well, after two decades of it, most of us found school to be second nature.

Our first didactic was a true intro course on what it means to be a psychiatrist, taught by Dr. Redding. We covered topics such as psychopathology (that is, the ways in which mental health breaks down) and forensics (the ways that psychiatry exists in the legal world). These courses covered the issue of treating patients without consent, to me the most gut-wrenching aspect of the field. It pains me to take away someone's rights by hospitalizing them against their will. But when a patient is actively threatening suicide, or so disorganized due to psychosis that he can't feed, clothe, or house himself, our mandate is very clear.

"Sometimes you have to be the bad guy for the patient's own good," Dr. Redding said in conclusion.

"I hate being the bad guy," I muttered.

"Oh, I love it," Rachel said nonchalantly.

The next didactic was led by a more junior faculty member who had just recently joined the teaching staff. Dr. Tony Strand (MD *and* PhD) had come over from our crosstown Harvard rivals at Mass General Hospital. Despite his impressive credentials, he turned out to be one of our most down-to-earth teachers and role models. He started out by telling us that his job was to try to get some psychopharm knowledge into us as noninvasively as he could while we continued to "do the heroic work of running the place behind the scenes."

It felt wonderful to actually hear someone acknowledge what we'd all been feeling—that we really were carrying the burdens of the hospital most of the day and night.

"The situation in modern psychopharmacology is tragically straightforward," he continued. "Most of what's done in the clinical world — the necessary, essential work of getting people better with drugs — is at best winging it off of a handful of anecdotal memories where something worked or didn't work. At worst, what goes on out there with the use of drugs on- and off-label is a travesty that has no evidence behind it whatsoever. Even your favorite, sparkling mentors on 4 South may fall into this trap now and again when the work demands action and there's no literature review available, there's no UpToDate entry, there's *nothing* to tell you what way to go."

"How can we become adequate psychiatrists if our mentors can't even rely on evidence-based protocols?" Erin asked.

"That's an interesting question. The answer will come with time and experience, and there are no shortcuts. My job here, though, is to teach you to embrace the skepticism you need in order to know when you're on track and when you're going astray. That's a skill set much more important than learning the specific doses and titration schedules in treating a particular disorder."

We looked at him eagerly, intently, waiting for the first lesson. We hadn't realized it, but at that point we were desperately thirsty for a bullshit detector.

"Okay, let's begin."

7

The Lightbulb Has to Want to Be Changed

By the end of the week, and with three mood-stabilizing medications on board, Deborah had settled back into a version of her usual self.

"I'm so ashamed," she said to me as she began to pack up her belongings.

"It's common for patients to feel that way once they're doing better."

"I was so out of control. The last time this happened, I said I would never let it happen again, but then —"

"It's a disease, Deborah. Sometimes it happens no matter how you feel or what you do."

"But I let it get out of hand. When it starts, it feels good — the hypomania. It feels like I can finally function. My mind is free, and I have the energy and the motivation to *do* things. It's hard to describe, but it feels so good, and then it just spirals out of control."

I felt like a fraud, once again, because I didn't know if the proper psychiatric intervention was to console her by telling her it wasn't her fault or to give her space to experience the guilt and shame that was taking hold. I noticed the clock in the corner of my vision: 12:30 p.m. I was late for court.

"I'm sorry, Deborah. I have to go. I'm proud of you and I think you're going to do well for a long time."

"Thank you for all of your help, Dr. Stern."

With a smile, I walked out and down the hall. "Court" was actually just a conference room at the end of the hall. Dr. Song said I could sit in for Jane's hearing. In the room there was a long table with some binders on it, surrounded by a handful of chairs. The judge sat on one side, while Jane and her attorney — a bookish woman in her mid-forties — along with Song and me lined up on the other.

"Let's get right into this. We are here to discuss the continuation request for court-mandated treatment on an inpatient setting for Jane West. The initial order is set to expire tomorrow at five p.m. I understand we have Dr. Song, who is the attending of record for Ms. West, here as an expert witness. Dr. Song, please provide a clinical update."

"Of course, Your Honor. As you can see, Ms. West's progress has been slow and limited by her ongoing refusal to eat regular meals. Her body-mass index remains at a concerningly low level, and I fear that without ongoing intensive treatment, the patient may begin to suffer irreversible physical harm from her malnourished state."

"And would the attorney for Ms. West like to respond?"

"Your Honor, my client has been admitted to this unit for seven weeks. As her chart reveals, she has engaged in regular group and individual therapy as well as adhering to her medication regimen without exception. She has become an exemplary patient in the eyes of the nursing staff, and I think we can all agree she has given this treatment center a chance. It is true that her weight has not risen as she would have liked by this time, but there are a number of reasons for this. First, as part of her condition, and as well documented by Dr. Song, eating has become ritualized, and these rituals simply cannot be undertaken in an environment such as this. Second, Ms. West indicates that the food here is not appealing to her and she would have more luck gaining weight at home or back at school. This brings me to the next point, that every week Ms. West remains an inpatient, she is away from her school, her friends, and her family — the very supports that may help her reestablish her academic status and re-attain the life she had prior to her illness. For these reasons, Ms. West is requesting a denial of the motion with subsequent discharge."

"And Dr. Song, given the criteria for ongoing court-ordered inpatient medical care related to imminent risk of self-harm or harm to others, can you comment on the patient's current status?"

"While Jane has not communicated any suicidal or homicidal thoughts or intentions, I am concerned that her body is not able to go on at this weight."

"But is the risk imminent?"

"I cannot say."

It felt like we were losing even though we weren't the ones whose fate was being decided.

The judge took a long breath and put his glasses on the table.

"Ms. West, I remain highly concerned for you. Strictly under the scope of my jurisdiction under the law, however, you do not currently meet criteria for court-mandated treatment at this time. I do not make this decision lightly, and I hope that you will continue to pursue aggressive outpatient treatment that is specialized in eating disorders. Do you understand?"

"Yes, Your Honor. Thank you."

As the judge adjourned us, Jane made eye contact with me as if to make sure I knew she had won. Won what? I wondered. She was going to die if she didn't give in and get the help she needed. It would break my heart if that happened. It seemed like she couldn't see that we were on the same side.

"Come with me," Song said as we left the room.

We exited the unit and headed for his office. It was a plain room, minimally decorated, with a bowl of mints at the center of his desk.

"Candy?" he asked, holding out the bowl.

"No, thank you."

"You are disappointed?" he asked.

I nodded. "She's going to die," I said.

"Perhaps. Probably not today."

"Doesn't it bother you that there's nothing we can do?"

"Nothing we can do? For seven weeks we have given her therapy, medications, a therapeutic milieu. We have given her food that she has rejected for seven weeks. The lightbulb has to want to be changed, Adam."

"Huh?"

"Oh you don't know that one? It's good. It goes like this. How many psychiatrists does it take to change a lightbulb?"

He waited.

"How many?" I finally asked back.

"Only one, but the lightbulb has to want to be changed, and it takes a very long time."

8

These Are Bad, but I Think You Can Do Worse

After several weeks and at the end of my first inpatient rotation on 4 South, I was beginning to take on the qualities of the patients I had been treating. I had started to notice the early warning signs of depression and anxiety. Sometimes I couldn't get myself to eat full meals. At times I felt hopeless and totally isolated. I had been spending every day surrounded by people in close quarters, but I felt more alone than at any other time in my life. I spent half my day as a reflective window for patients in their darkest hour. Some of their despair would be reflected back to them, and the rest would seem to come through and land on me. The days were getting shorter, and at night I would take the train home to my mostly empty apartment and feed blueberries to Magoo, the brown-and-white guinea pig I had adopted during medical school when I was still with my ex-girlfriend Eliana. I was skeptical of intentionally bringing a rodent into my home at first, but it turned out that Magoo was an especially good listener, particularly if I was feeding her blueberries. I was quietly relieved when I got custody of the little fluff ball after splitting up with Eliana. I was so starved for connection that

I found myself confiding in the guinea pig. I even began running cases by her, knowing that HIPAA's privacy rules applied exclusively to human beings. I presented my decisions and clinical care with self-doubt, but Magoo never seemed bothered, and somehow it helped my emotional state to give voice to what I had been struggling through during daylight hours.

Sometimes I needed more bidirectional interaction, though. Often I began an email or a text to Eliana only to delete it before sending. I would notice the intense draw of her warmth and our well-earned familiarity together, and it felt like a drug. I began asking myself why we had broken up. Was it because we weren't right for each other, or was it only that the timing wasn't right? I knew messaging her would be selfish because I always felt the urge to reach out the most when I was lonely. Sometimes, though, she would touch base with me, and I relished it.

Hey, she texted.

me: Hi

Eliana: can I tell you something quickly?

me: sure

Eliana: I have a guinea pig picked out to go on top of my christmas tree instead of a star.

me: ☺

Eliana: That was it. Pat Magoo for me.

me: Will do

When we did talk on the phone, we used our shared love for Magoo as a benign excuse to check in.

"How is fluffy Magoo?"

"Pretty Gooberific," I replied. "But poor Goobs had to spend five hours in the room with my beeping pager yesterday. I had left it at home, and it was going off the entire day nonstop. All I could think of were her poor tiny ears." I paused. "They're quite large, actually, as you know."

"I bet she thought it was exciting! Her velvet ears don't get much excitement."

"That's true," I said.

"Anyway, I should get to sleep. I just wanted to check in and talk about fluffy guinea pig ears."

"Good night, Eliana."

The chats always left me feeling briefly lifted, but then often my mind would go down rabbit holes wondering if I should have only applied to residency programs in New York. If I were in New York, I would have been close to my family and Eliana, and I probably would be living a very different life, full of social connection. These thoughts brought me to a dark place. I had to deliberately remind myself that Eliana and I had given it our best shot and still chose to break up and that I had ranked Harvard high on my list, for good reason. From Longwood, the opportunities were endless, and I could become the psychiatrist and the man I wanted to be.

Another emotional outlet was, unsurprisingly, Feelings class, which had gained momentum as we became increasingly ensnared in the world of psychiatric training over our first several months together. Nina led the class with a kind of radical acceptance, knowing that we were all going to mess up constantly and that as a group we would put ourselves back together. Her openness encouraged something beyond honesty. It was almost like each class started with a competition to see who'd fucked up the most in the last week but ended with us all having learned from one another.

Miranda kicked us off.

"So, this is really embarrassing."

Nina nodded as if to say, *Enough apologizing. Get to it.*

"I was interviewing a patient with delusional disorder who had just been admitted the day before. He has this fixed belief that his mother is spying on him with hidden cameras throughout his home. Everything else about this guy seems totally normal."

"Is there any chance his mother is spying on him?" Erin asked.

"I wondered that too," Miranda replied. "But I asked his wife about it, and she said his mother is eighty-five years old and living in a nursing home after having a stroke. Either it's a very elaborate ruse or he's delusional. He doesn't see it, like at all, though. He is completely convinced that she's doing this."

"It would be interesting to know when the delusion started," Nina said. "It wouldn't surprise me if it coincided in some way with his mother's decline."

It was the kind of grown-up psychiatric interpretation that we all aspired to.

"I will look into that," Miranda replied. "In the meantime, he got admitted yesterday and at our first meeting he began telling me that he didn't trust me because his mother had obviously already gotten to me. And why did he think that? Because he saw that workers were installing cameras on the unit. The attending jumped in and said, 'Sir, I assure you that we are not installing cameras here.' We finished up the interview and stepped outside the exam room, where I saw a crew of three guys up on ladders installing security cameras at the door! The patient just looked at me with disgust and walked back to his room. I could have died, I was so mortified."

"At least you only could have died," Rachel said. "A patient came up to me in the hallway and told me they *were* dead and asked if I could help find their body."

"Oh my God," Miranda said.

"Cotard's delusion," Nina announced. "Not that common, but it happens."

"What did you do?" Miranda asked.

"I had no idea what to do, so I said, 'Okay, that sounds really difficult,' and stared at her for a few seconds until the next group session started and I was saved from having to know what to do."

"What I like about your reaction, Rachel, is that you took her concerns seriously and validated her distress. Next time you see her, try asking her what it's like to feel like she's dead and just listen for a while."

As the class wound down, I started to dread going home with only a guinea pig to share the burden of my neurotic self-doubt.

"Hey, Rach. What do you have on deck for this weekend? Want to do something?"

"I'm starting my medicine rotation on Monday, and there's a bunch of getting-to-know-you stuff scheduled. I hate getting to know people."

"Oh, okay. Miranda, how about you?" I asked.

"Getting together with my family. Actually, it's this very

cool place that we go to each year for a kind of reunion in Pennsyl—"

"Erin?"

"I'm trying to get Bobby to take me on a romantic weekend. He's been so miserable since we got here, and I'm stuck at work so much. I just need to spend some time prioritizing him this weekend."

To me it seemed like she always prioritized him, but I'm just the guy she sat with every day for hours at a time. What did I know?

Sensing I was going to spend another weekend talking to Magoo, I opened the dating app on my phone and started scrolling.

"What's that?" Rachel asked. "Are you on a dating site?"

"No," I lied.

"Yeah, you are. Let me see your profile."

"Absolutely not."

"Come on."

"Not a chance."

"Fine, I'll just find you on the site," she said, exasperated. "You could at least tell me about the bad dates you go on."

"Whatever you say, Rach."

Over the next few weeks, I managed to have three certifiably disastrous first dates. The first woman was in training to be a massage therapist and was stunningly beautiful, but she led off by telling me that she didn't "believe in Western medicine." We clearly had irreconcilable differences. I texted Rach immediately when I got out.

"Next!" she replied.

I followed up with texts about a habitual spitter and a woman who didn't speak more than one word at a time.

"These are bad," Rachel replied, "but I think you can do worse."

The whole dynamic with Rachel was odd but kind of exciting. At least she was taking an interest in me, and while the dates were not going well, it did allow me to kind of feel like the whole thing was a joke and Rach and I were the only ones in on it.

Then something unexpected happened.

I met Ashley at the bar in the lobby of the Four Seasons right off the Common. She was a senior in college and couldn't wait to graduate and enter the real world, so it felt like a good strategy to take her someplace upscale. We met and each breathed a sigh of relief that the other seemed close to how we'd presented ourselves on the website. Then we ordered fourteen-dollar drinks and found seats that were set too far apart. I found myself shouting at her and then not hearing her responses. I was beginning to lose hope and had already started mentally preparing my text to Rachel when Ashley surprised me with a good idea.

"I know we just got here, but it's a little hard to talk. Do you want to go for a walk outside?"

I nodded.

We made our way to the Boston Common and continued into the Public Garden. It was freezing out, so we were pretty much the only ones there, which felt very romantic whenever I could stop breathing into my cupped hands to prevent frostbite. I would blow and look at her, incredulous. On paper and in person she was pretty much perfect—interesting and clearly smarter than I was, but also engaging and kind. I had learned in our pre-date chats that she came from a family similar to mine, except they were Yankee fans instead of Mets fans—a forgivable sin under the circumstances, though some might disagree. I looked at her walking through the park and realized I didn't know what to do with perfect. Luckily, she was better at first dates than I was and started asking me questions.

"What is it like to be a psychiatrist?" she asked.

"Psychiatry resident," I said, clarifying. "It's pretty surreal, actually."

"How so?"

"It's like this. You dream for half your life about becoming this ideal. You devote yourself to it. You sacrifice for it. And then when you finally get there, it doesn't feel the way you imagined it and you realize that it was never going to."

"Why not?" she asked.

"Because you're a doctor, a psychiatrist, in the real world, and the notion I was carrying around for ten years was a fantasy."

"It sounds like it's been a letdown."

"Not exactly. The highs can be so high, when it feels like I've actually helped someone. It's just that I don't get to do that as often as I thought I would. There's a lot more scut work—"

She looked at me with her eyebrows raised.

"Paperwork, busywork nonsense that falls to me as the intern, and half the time I'm being tasked with treating someone against their will or locking them up when I never wanted that to be my job. It's a lot of responsibility, and sometimes I just feel like it's too much weight for one person to carry."

I heard my litany of complaints about work and thought how pathetic I must sound.

"I'm sorry," I said. "I don't mean to whine."

I looked down at her and met her eyes with mine. They didn't seem frightened or bored; they looked engaged. And then they were closed and she was leaning into me. Our lips met, freezing me in place. My mind went blank for two seconds, and as she pulled back and we locked eyes again, she exhaled, and I could see her breath take shape and then vanish in the frigid air.

"I don't think you always have to be alone with all that weight."

I smiled. It was all I could think to do. She took my hand, mitten to glove, and we kept walking through the barren garden.

me: i just got home from another date

me: it was great

Rachel: with the girl from MIT?

Rachel: or the nurse practitioner

me: yeah — Ashley, the girl from MIT. the NP got back to me that she's working like crazy and how about next weekend.

Rachel: boo

me: good point

me: she turns out to be one of these people that is so "good" that she drives you crazy. i don't think it would ever work.

Rachel: oh god

me: her life is a series of travels devoted to helping HIV pts in Africa, and what does she like to do for fun? she likes to run.

Rachel: bad news bears

Rachel: so idealistic

Rachel: barf

me: don't vomit. it's not polite.

Rachel: i am not polite

Rachel: so it's fitting

Rachel: so are you going to ask her out again or what

me: i do not know. i've got to let it simmer a bit. I had a better time with Ashley. She seemed great! That's what I was trying to tell you.

Rachel: blah

9

Night Float

The timing of my new romance was awful. I was entering night float — a rotation that involved working two consecutive weeks in a row completely at night. That rotation would be immediately followed by Ashley's long winter break. I remembered how wonderfully mundane and restorative it was as a college student to have an entire month off in the middle of winter and wondered how I had come so far from that life in such a short time. Medical schools offer one last summer break after the first year and then no extended breaks for the remaining three years. Then residency treats time off like any other job but with more restrictions because *someone* has to be in the hospital to take care of patients at all times. An individual resident's vacations are always highly coordinated with other members of the class and at the discretion of the program overall.

Over the next two weeks the person staying with all of the hospital's psychiatric patients overnight would be me, along with the second-year resident, Rebecca. We would start service at six o'clock each night and work nonstop until the sun rose, when we would begin to prepare our verbal sign-out for the day team. This was supposed to occur by 8 a.m., and then we were to go home and try to sleep in broad daylight before starting the whole process again. There was no rest for the weary on night float. I would soon learn that with all of the loose ends to tie up and the complexities of the cases to convey that getting out by eight was a fantasy. After a particularly quiet night, I could sometimes make it home by ten, leaving only eight hours until I was due back at the hospital, showered and ready to carry the load of the entire day team once again. It was an unrealistic race with the clock to fall asleep as fast as I could, leaving no time for any social activities at all. I had only been able to see her a second time the night before the new rotation started, and I soon discovered that even finding times to text with Ashley proved to be difficult and seeing her would be impossible. Worse still, I knew that by the time my schedule returned me to daylight, Ashley would be off at home in Arizona until the end of January. The early embers of a relationship needed to be stoked. But how could they be?

As I walked onto 4 South on the first Sunday night, Rachel was waiting to give me sign-out.

"How bad is it?" I asked.

"Up here is fine, but downstairs is a mess," she said, referring to the Green Zone of the ER, where patients needing psychiatric services were located. "But really what I want to know is how bad was your date last night?"

"It was fine — good again, actually."

"Are you sure?"

"Yeah, we had a nice time."

"Stop. Not interested anymore."

I felt the blood start to gather in my chest and face. Sometimes I found Rachel to be infuriating. Her patients always seemed to find a structured and casually empathic calm in her therapeutic encounters, but she had a tendency to make my blood pressure rise with her casual dismissiveness of me.

Over the course of her giving me sign-out on each of the patients, my pager went off three times. It was Rebecca from the Green Zone.

Need assistance. Come down after sign-out's done.
~Rebecca 17:55

Seriously. Situation deteriorating rapidly. Come down STAT.
~R 18:03

SOS 911. ~R 18:09

"Well, I better get down there."

"Good luck. You're going to need it."

I stopped for two croissants and two coffees in the lobby coffee shop on my way to the ER. I typed in the five-digit code to get into "the bunker," a tiny room with no windows that the psychiatry staff holed up in to get work done when near the emergency room. There I found Rebecca staring blankly at the whiteboard that listed all of the patients waiting to be seen.

"Breakfast?" I asked.

"You stopped for coffee? There's no time for coffee. Look at this. Look."

There were eight patients waiting to be seen.

"I don't understand how they can have a whole team of residents and attendings during the day and leave one second-year resident to carry the load all night."

"Well, and an intern," I added.

She shot me a look.

"Right. And you. Well, we better divide and conquer. You take these four, and don't send anyone home without first presenting the case to me. Understood?"

"Aye-aye."

But she was already out the door with her clipboard. I sighed and took a slug of my coffee, looking up at the names of the four patients Rebecca had assigned me.

"What the fuck?" I asked.

It was practically my whole patient list from the month before. Jane, Paul, Ginger, and Deborah were all back in the ER within a matter of weeks.

"Impossible."

"What's impossible?" a woman from the other side of the room asked without looking up from her computer.

"Oh, nothing. I'm just surprised to see some of these names."

I turned my gaze from board to woman stationed across the room.

"You've been sitting there this whole time?"

She nodded.

"Who are you?"

"I'm Nancy, the bed finder. I find all these patients beds either here or around the city when they're admitted. You're going to want to be nice to me because I'm the only way these patients get out of your ER."

"Good to know, Nancy. Want a croissant and a coffee?"

"Well, thank you very much," she said, sticking out her hand, still not looking away from her screen. "So what's impossible?"

"It's just, this list of patients I've been assigned, they're—they're all the patients I was taking care of up on 4 South."

"So what?"

"Well, they were doing well when I discharged them. I guess I'm just surprised to see them back here."

"Kid, they always bounce back."

I took a bite of my croissant and headed out with a clipboard I found next to the printer.

The recent court victor, Jane, was the first patient on my list and the first room I came across on my way through the Green Zone. Like all of the patients in this part of the emergency department, she was being held behind glass sliding doors in a barren room with only a gurney inside. Her mother sat next to her with puffy eyes staring at the floor.

"Jane," I announced as I entered.

Impossibly, she looked thinner and more frail than when she had left.

"Oh good, it's you."

"It's me."

Our eyes met as we took each other's measure.

"What brings you into the emergency department today?" I asked, knowing exactly what the answer was.

"Are you an idiot? Do you need me to tell you everything all over again?"

"I find it's just good to let my patients tell me what's going on."

"I'm not your patient. I'm just seeing you because you're the only idiot here."

"Okay, Jane. So, help an idiot out, will you? Up on 4 South, you told me how it was all going to go, and you were right. What about now?"

"Times have changed. This time Medicine and Psychiatry are going to have a pissing contest over whether I'm stable enough to be admitted to Psychiatry. Medicine's going to say I am fine to be admitted to 4 South, and Psychiatry's going to say my body-mass index is too low, along with my heart rate and glucose."

"Who's going to win?"

"Psychiatry, of course. They never take anyone that makes them nervous medically, and I make people nervous."

"Okay, Jane. Thanks for showing me the way. Anything else I should know?"

This time I made contact with her mother, who just shook her head.

"Your shoes are untied."

I looked down. Right again, Jane. I bent down to tie them and looked back up at her.

"I'll let you get some rest, but give a shout if I can do anything for you."

"Let me get some rest?"

She was adept at calling out nonsense.

"You know what I mean," I said, exiting and closing the sliding glass door.

Next up was Paul, our unit Romeo whose girlfriend had moved straight from 4 South to Sweden. When I entered Paul's room, he told me he was suicidal again. I asked if he was still heartbroken, and he said he was.

"It's incredibly hard when someone you care about moves away," I said.

"She didn't move away this time."

"This time?"

"There's someone new. Well, there was. She dumped me, though."

Paul had apparently rebounded hard after his discharge from 4 South, and I was beginning to scan through the *Diagnostic and Statistical Manual*'s list of personality disorders in my mind. He didn't quite fit with borderline personality or narcissistic personality, but one could make an argument for dependent personality in someone who repeatedly falls so hard and fast that they become suicidal when it doesn't work out.

I asked him about his suicide plans, and he listed off seven or eight potential methods of ending his own life. With that, I knew that Paul had just qualified for another inpatient stay or at the very least that I could not discharge him. Rebecca had taught me early on that ER psychiatry is a different kind of animal from all the other environments we rotate through.

"The key is that every encounter ultimately boils down to will you admit the patient or will you discharge them," Rebecca said. "You can't do therapy in the ER. You can't give medicine that will make much of a difference. All you can do is decide where the next treatment is going to happen."

"What about a therapeutic encounter? Empathic listening? Can't you make a difference that way?" I asked.

"Oh, yeah. Of course that. If you have time," she replied without any hint of irony or sarcasm.

Next, I saw Deborah, who had been looking really well by the time she was discharged after the manic episode on 4 South. Her room was dark, with the curtains drawn in front of the glass.

"Deborah? It's Dr. Stern. May I come in?"

I heard a muffled response and poked my head in. Her face was buried in the pillow. I sat down next to the bed and waited for her to engage, but she didn't.

"Deborah, can we talk?"

Slowly, she turned her face to look at me.

"I can't live like this. I'm taking the medicine, and it's just making me more depressed every day. Why is this happening?"

She looked to me for guidance, but all I had was empathy.

"I'm sorry this is happening to you. It's the disease."

"What am I going to do?" she asked desperately.

"We're going to take care of you," I replied.

There were mounds of paperwork beginning to pile up for each of the three patients I had seen, but I was determined to see the last of the four I had been assigned before returning to the bunker to work on my notes. The fourth patient was Ginger, whom I had previously helped with a small mirror. When I entered the room, though, it occurred to me that all of her possessions including the mirror had been stowed away by security. She was pacing and self-dialoguing in a way that I couldn't follow. I couldn't get her to look at me or even say hello. I stepped out of the room and asked around until I was able to find another small pocket mirror. I held it out in my hand, which caused Ginger to stop pacing and look up at me for the first time. She reached out and gently took the mirror from me. She looked at herself in it, and I saw her face tighten and her lips curl down. She threw it down onto the tiled floor, and it cracked. Then she looked up at me, her face still in a grimace, and began to pace again.

"I'm sorry you're having a difficult time," I said as I scooped up the broken mirror.

I stepped out of the room and told the nurse on duty I'd be ordering antipsychotics for her momentarily.

When I got back to the bunker, Rebecca seemed even more frazzled than she had at the beginning of the shift.

"Nobody's doing very well, I'm afraid," I said by way of report.

"Write the notes and move on," she replied. "Four new consults have already come in."

10

Controlled Chaos

Working nights with Rebecca, I felt my emotional state quickly descend into a kind of controlled chaos. The work was intense but required a delicate approach. It was definitely harried, but every patient needed to feel heard. There was never enough time. Even when I was sleeping, my dreams always seemed to run out of time. Unconscious stories were left half-finished and bled into the eerie world of my nighttime working hours.

It should have been no surprise that my attempts to keep something going with Ashley fell flat. We were on opposite schedules, and every waking hour there was something I needed to be attending to. I started out texting her on my way to work, but if I didn't catch her at just the right time, we would end up having awfully drawn-out conversations bit by bit over days. Eventually, I told her that work was really intense and, if it was okay, I'd reach out when the two weeks of nights were finished.

On the fourth night of the rotation, I encountered Beatrice. Rebecca received a page, and I saw the look in her eyes — as if she would implode like a neutron star if she had to return the page with a phone call.

"I'll take this one," I said heroically.

"No, it's okay. It's just Beatrice paging the emergency pager from home."

"Patients can do that?"

"They certainly can. All they have to do is call the hospital operator and ask for the psychiatry resident to be paged to their number. You'll get to know her well next year."

"What do you mean?"

"You haven't heard about Beatrice?"

Rebecca's face brightened.

"Nope."

"I see. Why don't you take this one, then."

I called the number and an elderly-sounding woman answered the phone.

"Hello, who's this?"

"Oh, hi, it's . . . uh . . . Dr. Stern. I'm the psychiatry resident on call tonight and —"

"Thank God you've returned my page."

"How can I help you this evening?"

"Well, I can't sleep."

"Okay. And what do you usually do when you can't sleep?"

"Well, Dr. B prescribes a sleep aid, but I've already taken it. So now what am I supposed to do?"

"I pulled up your chart, and it says he prescribes zolpidem five milligrams for you. When did you take it?"

"At eleven p.m."

"Ma'am, that was — well, that was just five minutes ago."

"Yes."

"Usually it takes longer than five minutes for these medications to kick in. Why don't you try to sleep for the next twenty or thirty minutes, and if you're still awake at eleven thirty or so, take a second tablet."

"Thank you, Doctor. You are wonderful."

"Okay, my pleasure. Good night. Sleep well."

I hung up and looked at Rebecca with shrugged shoulders.

"Not so bad. Seems like a sweet old lady to me."

Rebecca started cackling. It was unnerving.

At 11:35 the next page came in.

"Look, Dr. Stern. It's for you!"

I called Beatrice back again.

"Hi there."

"Oh, thank God. Dr. Stern, I still cannot sleep."

"I'm really sorry to hear that. Did you try taking your second tablet?"

"Well, no. I remember you saying that, but I just didn't feel comfortable doing it without confirming that you meant a second tablet of the zolpidem and not one of my other meds."

"Yes, of course I meant — I'm sorry, Beatrice. Yes, take another zolpidem, and I will put in a message to your doctor to let him know that he should consider adjusting your regimen."

"Okay, thank you, Dr. Stern. God bless you."

I hung up.

"Hadn't taken the second dose, but she will now," I said to Rebecca.

"I'm sure that will be the end of that and you won't hear from her again," she said, her voice dripping with sarcasm.

I heard from her twice more that evening. The second time, after 1 a.m. at that point, I responded differently.

"What seems to be the emergency, ma'am?" I asked angrily.

"Well, to be sure, it all started several years ago when —"

"I'm sorry, Beatrice. This is a pager for emergencies only. Can you tell me what the emergency is?"

"I can't sleep. Haven't you been paying attention?"

"That's not an emergency."

"It is to me."

"How often does it happen?"

"Every night."

"Then it's just your evening routine."

"Dr. Stern, I do not appreciate your tone. I thought you were different."

"Please only use this pager for emergencies, ma'am. Good evening."

I realized I was standing over the base of the phone in an almost domineering manner and collapsed back into the chair. Rebecca put a hand on my shoulder.

"In my class, we call her the Page Torturer. It's like her superpower."

"That's mean. She doesn't do it on purpose, right?"

"Doesn't matter," she replied.

"I feel bad. I thought maybe I was different, but she's right. I'm not."

"Well, the good news is that you'll get to have another shot at it tomorrow, and then the next night, and pretty much every time you're on call until you graduate."

The pager buzzed again.

"Don't tell me it's her again."

"No," she replied. "But it's one you saw on our first night here. The patient with anorexia who got admitted to Medicine. The primary team wants someone to come see her."

"What, now? It's the middle of the night."

"It's always the middle of the night. What's the problem?"

"I just don't feel like being berated again tonight. I'm never any match for Jane."

"Instead of trying to match up with her, maybe you could try aligning with her. Off you go."

With a sigh, I exited the bunker, walked out of the Green Zone and past the Trauma Zone to the ER exit. I took the elevator off the lobby to a sky bridge that led to another part of the hospital. For just a moment I pressed my forehead against the glass that overlooked the city. It seemed so quiet. I could almost see my apartment amongst the brownstones and imagined how good it would feel to get into my own bed when it was dark out. My fantasy ended as an ambulance with flashing lights made its way underneath. Duty called.

I took the elevator up to the eleventh floor, where Jane was admitted. I entered the room with a knock and seemed to startle the one-to-one sitter, an employee who had been ordered to sit with Jane because of her tendency to pull out the nasogastric tube that was feeding her. I had learned that since our interactions on 4 South, a different judge had approved an emergency order for tube feeds due to the imminent risk of death from malnourishment if she did not receive them. Still, Jane fought.

"Hi," I said quietly. "Is it okay if I sit down for a bit?"

"When do I ever get what I want?" she retorted.

I told the staffer that she could go for a while and that I would stay with Jane at least until she got back.

I pulled up a chair almost parallel to the hospital bed, so we were both staring at the blank TV hanging from the wall. We sat quietly for a full minute or two. The silence felt uncomfortable at first, but eventually it instilled in us a kind of calm. Our breathing was slow, and we were just still and at ease. Eventually, Jane broke the silence.

"I asked them to call you, you know."

I shook my head.

"I needed to talk to someone, and you're not as bad at it as most of the people around here."

"What's on your mind?" I asked.

She motioned to an even more serene view of the city I had gazed upon just minutes earlier.

"Do you ever look out there and wonder if you'll ever have the kind of life all of those people get to have? Sometimes when I'm awake and it's late, I start imagining my friends and my classmates—well, my former classmates—all out there sleeping in their beds dreaming their little dreams. Tomorrow they'll wake up and maybe they'll be inspired to go do something or be something or achieve something. Wouldn't that be wild? To have a life that could support your dreams? To have control over what you were and what you could become?"

"I do have thoughts like that," I admitted.

"What do you wish you had control over?"

I had never really thought about it.

"That's a really good question," I replied. "For one, I wish I always knew how to help my patients. Sometimes I do, but often I don't."

"You're an intern," she said. "You'll get better."

I had learned in didactics that the prescribed maneuver when the conversation turns to the psychiatrist is to try to reflect the topic back to the patient and probe for emotional significance.

"Do you feel let down sometimes because it seems like no one can help you?"

She shook her head, and tears began to well up in her eyes.

"I feel like I should be able to help myself. I know what everyone is thinking. *Just eat,* but I can't do it. Not consistently."

She paused. I had no words. My lack of experience was showing again.

"I don't want to die," she said quietly.

All of the responses my mind went to were wrong. *You're going to be okay. We'll find a way through this. It's all right.* In real life, if Jane were a friend, I'd give her a hug, but that couldn't be the way of a psychiatrist.

I leaned forward and connected with her eyes.

"I know," I said.

That was all I had, but in the moment it was enough. We sat quietly together for a long time, and eventually my pager went off.

"Go," she said. "It's okay."

Turning my back to her and leaving the room felt wrong in my gut, but still I did it.

11

Insomnia and Electric Shocks

When my two weeks of first-year night float mercifully ended, I became profoundly uneasy with living in the real world again. How could I simply go back to normal life when I knew what happened at night, every night, ad infinitum? Patients come in. They are evaluated. They get placed. They get treated. They get discharged. Rinse and repeat. Does anyone ever get better? It felt like I had seen behind the curtain, and what was revealed was more than I was ready for.

My fucked-up sleep cycle didn't help matters. I forced myself to stay awake during the day and then couldn't sleep a wink at night. On my fourth consecutive hour of staring at the ceiling one night, I decided I needed to break the cycle. I had to escape my bedroom, which had begun to feel like a prison cell. I stood up, put on the first clothes within reach, and walked out the door of my apartment.

The streets were eerily quiet. Sometimes, being surrounded by all of the chaos inside the locked door of the inpatient unit, I lost sight of how calm Boston could be when it was at rest. I found the empty streets comforting, and even felt my heart rate slow. Passing by brownstone after brownstone along Commonwealth Avenue, I wondered what kind of dreams the sleeping people all around me were having.

Then I began to wonder about the patients at the medical center. Were they sleeping, or were they as tormented by the night as I was? Without my deciding what I was going to do when I got there, my steps led me the mile and a half back to the hospital. I used my ID badge, which finally no longer read PHYSCHIATRY after being fixed, to get into the building, around the corner, up the elevator. Like so many on the unit, I couldn't fully comprehend the chain of events that had led me there, but I swiped into 4 South and it felt like relief. I was in sweats and a hoodie, but the checks person didn't flinch.

"Dr. Adam."

"Hey, Reg."

I walked past him casually and into the nurses' station, where no one was bothered by my appearance. I looked up at the whiteboard that contained the patient census and found Ginger's name. I poked my head out of the nurses' station, and there she was walking the halls. She was talking to herself, but somehow she already seemed much more at ease than in the ER. I wondered if Ginger actually felt more comfortable on 4 South than anywhere else in the world.

I stepped out into the long hallway down which she paced. As she passed, I sought to make eye contact with her, but her gaze was fixed on the floor in front of her. I was about to say hello, but I thought it might be more disruptive than helpful, so I refrained. Instead, I exited the unit as quietly and unremarkably as I had entered. I took a cab home and fell asleep almost the instant my head hit the pillow.

• • •

The next day, when I arrived back for daytime work on 4 South, Dr. Song stopped me in the hallway.

"I need to see you in my office," he said sternly.

I followed him out of the unit around the corner and into his office.

"Mint?"

"No, thank you though."

"I heard you were on the unit last night."

He waited for a response, but I sat frozen in the chair. I had been noticed after all.

"It was your night off, but you came to the unit. I have seen this before. It is never good, Adam."

I gave a slight nod.

"Were you checking up on someone?"

"There were a few patients I saw in the ER that I thought may have been admitted to 4 South."

"And what were you hoping to achieve by coming here?"

"It wasn't a very well-thought-out plan."

"More of an instinct. Yes," he said. "Let me give you an essential piece of advice and a clinical directive that I expect you to *try* to follow."

"Okay," I replied hesitantly.

"When you are here, you work hard and do what needs to be done for your patients. I know this is already true for you, so it will not be difficult. When you leave here, I suspect your mind cannot leave the work here. It travels with you. I need you to start constructing a barrier between your life outside of these walls and within them. Do you understand?"

"I think so."

"If you do not separate these worlds enough, you'll burn out, and you have not been afforded the luxury of burning out. You are needed now and in the future."

I nodded again.

"No more three a.m. visits, Dr. Stern."

"Understood."

"Let's go do rounds. You can take over care for Ginger and Deborah if you'd like."

"That would be great."

"If we adjust rounds, maybe you can observe Deborah's first ECT session this morning."

"She's getting ECT? Today?"

The idea of electroconvulsive therapy frightened me, but the truth is, I had never seen it.

"I have brought it up with her a number of times before, but the negative associations people have with this treatment are profound. This time, she has agreed to try it."

"Why now?"

"Not clear, though she said she would do it specifically for her family. I think it will be very good for her and for you to observe. Now let's go."

When it was time for Deborah to go to ECT, she and I waited at the unit door. Flatly, Deborah stared down at the tiled floor. The fiery energy that engulfed her during the last manic episode was just a memory, leaving an enormous void in its departure. People often misunderstand depression as a kind of intense sadness, but more often it's an unbearable absence of feeling that patients describe. Deborah had told me earlier that morning that she just felt entirely numb.

When the transport person arrived at the unit with a wheelchair, Deborah said she didn't need it.

"It's for your trip back, dear," Reg replied from three feet behind us.

It was the kind of foreboding statement that spooks people about ECT. *What is going to happen to me after they course electrical current through my brain?*

The transport person walked us in silence down the hall to an elevator that led to the ECT suite a couple of floors below. She used a badge to unlock the door and told us she'd be back to pick Deborah up as soon as she was finished.

We entered a plain room with faux-wooden flooring and curtains that draped between several patient stations. A smiling nurse escorted Deborah to a restroom, where she was to empty her bladder so as to avoid the risk of involuntary urination during the procedure. Dr. Macy introduced himself to me. He looked to be about my parents' age, with graying hair on the sides and back. He was wearing a bright tie and checkered suit. I had thought he'd be in scrubs.

He asked me if it was my first time observing ECT.

"Yes, it is. I don't really know what to expect."

"Well, don't get your hopes up for a good show. The whole procedure is over in a few minutes, and it's all pretty boring. I would know — I've been doing it for thirty-some-odd years now."

"Wow. How did you get into it?"

"In the eighties. I was a junior faculty member, and the department chief at the time just sort of said, 'Hey, we've got a need for you there,' and thirty years later I'm still here. Kind of different than how they treat young people today. Everybody gets to *pick* a career now. Do you know what you want to do in psychiatry?"

I didn't have the faintest idea, but I didn't want to seem aimless, so I said I was thinking about child and adolescent psychiatry.

"Well," he said after a pause, "let me give you my little spiel about all this. ECT has been around for eighty years, and we still don't know exactly why it works. Like a lot of things in psychiatry, it was discovered somewhat by accident. There are a lot of theories about why it works that tend to invoke the concepts of neurogenesis, brain plasticity, and rapid global neurotransmitter release. If we're being honest, it's still all a work in progress. While we don't know *how* exactly it works, we definitely *do* know that it *does* work. Seventy to eighty percent of patients with mood disorders like depression respond very well to this treatment. That's about twice as good as most antidepressants or therapy."

"Wow. So why do you think more patients don't get the treatment?" I asked.

He sighed.

"Well, there's a big stigma associated with ECT. Psychiatry comes by stigma quite honestly, if we're telling the truth. We work in the field that once employed frontal lobotomies, remember. *One Flew Over the Cuckoo's Nest* didn't do us any favors. But like I said, ECT is pretty mundane. The anesthesiologist helps the patient go to sleep and gives them a drug that prevents them from convulsing. Then we administer the treatment. The seizure is almost always less than a minute long, and then there's a recovery period, where patients are sort of out of it. Look at Reynaldo over there."

I turned toward a patient in bay 2 who had already had his session. He was mindlessly scrolling on his phone.

"Is he registering anything, or is that entirely automatic?" I asked.

"Probably automatic. The treatment usually leaves patients unable to encode new memories for a while, but that part wears off eventually. Every patient is different. That's why, like with your patient today, we'll find the seizure threshold for her in particular and then increase the intensity based off that in the future. Tell me a bit about Deborah."

"Oh, well, she's a middle-aged woman with bipolar depression. She was manic just recently. Seems to have crashed pretty hard after we got her down from that."

"Yeah, I know all of that. Patients get a consult before they come down here. I mean, tell me about her. Any family? What are the stressors in her life?"

I was ashamed to realize that I didn't know the answers with any depth.

"She definitely blames herself for her illness," I said, deflecting.

"Always find out about the people behind your diagnoses. That's the most important part of this whole deal," he said, motioning broadly to the room around him.

It was unexpected that the guy who worked with patients under anesthesia would advise me on the importance of knowing about patients' lives, but Macy seemed to know what he was doing, and I had learned to listen when someone tells me something they think is important.

Deborah came around a curtain, and the ECT nurse led her to bay 3. She sat down hesitantly, and stone-faced.

"Hi Deborah, I'm Dr. Macy. I'll be the doctor performing your ECT sessions."

She nodded slightly.

"It's common for patients to feel nervous before their first treatment because they don't know what to expect. Let me tell you what's going to happen. First, my colleague from anesthesiology here — say hi, Mark —"

The anesthesiologist waved and smiled as he arranged his gear to set up an intravenous line.

"That's Mark. He's one of the best, if not *the* best anesthesiologist in the hospital. He's going to get you going with an IV and some medications that will help you relax and, when we're ready, go off to sleep. There's the methohexital that puts you under, and then there's the succinylcholine, which prevents your muscles from convulsing during the treatment. They're drugs we use all the time, and they're very effective and safe. Any questions so far?"

She shook her head.

"Then Dr. Stern and I will set up the ECT device, and when we're all ready and you're totally under, we'll administer the stimulation, which goes from one of these paddles to the other."

He held up the two ECT electrodes.

"In your brain, there will be a brief seizure happening, but out here it will all be very quiet. Only your foot will shake back and forth because we'll have a blood pressure cuff up to prevent the meds from getting to that foot. We do that just so we have a visual sense of how long the seizure lasts. The device also gives us information about what's happening inside, which is helpful. The seizure will last for under a minute usually, and then you'll begin to wake up over the next ten to twenty minutes. By the forty-five-minute mark, you should be up and on your way back to 4 South. Any questions yet?"

She shook her head again.

"All right, we're going to take good care of you," he said as he turned to face the machine.

"I'm glad you're here," she mouthed to me.

I smiled back at her.

"Hey, I need to ask you something. I'm kind of embarrassed I never asked you before. Who are the important people in your life?"

She seemed almost taken aback at first.

"Well, there are my twins. Fourteen-year-old boys. They live with their dad now though. We divorced last year, and he got majority custody. I think they didn't like that I had bipolar disorder, but when you go through that kind of experience you realize how little in this world you can control. I think that sometimes discrimination is just the price we pay for existence."

She was getting choked up and stopped there.

I felt awful realizing I had treated her throughout her stay on 4 South and never knew about her children. I wondered if the job was draining me of some aspects of my humanity.

"Do you think this is going to help me?" she asked.

"I do. I really do."

"We're ready," Macy called out.

"Okay, I'll see you afterward," I said to Deborah.

"Oh, good. See you then."

She seemed relieved.

Mark placed the IV effortlessly and began to give her the sedating medicine. Within a few minutes she was under, and just after that it was our turn to perform the treatment.

"It's all yours, Dr. Stern," Macy announced. "Remember, we're titrating the intensity, so it may take us a few attempts to get an adequate seizure."

The nurse held the paddles in place, and a green light appeared on the device indicating a good connection was being made.

"Delivering stimulus," I said awkwardly.

What is one supposed to say when they are about to send electricity into someone else's brain?

The device held a loud tone for two seconds and stopped. Nothing happened.

"Did it work?" I asked.

Macy shook his head. He motioned to the right foot with the blood pressure cuff tourniquet applied.

"That foot will start shaking when it works. Raise it to three point five seconds."

"Delivering stimulus."

The tone came on and lasted just a bit longer. This time Deborah's body tensed for a moment before relaxing. Her foot still didn't shake.

"Again, but up to five seconds."

I adjusted the knob on the machine once again.

"Delivering stimulus."

On that third attempt her body tensed then relaxed, and at first there was nothing and my heart began to sink. After another second, though, I saw that right foot begin to twitch rhythmically and I let out a big sigh of relief. The seizure lasted twenty-five seconds, and Macy showed me confirmation from the crude electroencephalogram that the device was printing out.

"Well done, Doctor," he said.

I looked back at Deborah, still unconscious and looking so peaceful in that moment, and wondered what was happening underneath it all. Maybe relief could still come for her.

12

Bread-and-Butter Medicine

With our schedule featuring overnight calls and then week-long periods of night float at multiple clinical sites, our class developed a rapport with cyclical depression. There was always a handful of us who were rotating through light rotations that only required daytime work, but there were also always a certain number of us who were either on call, post-call, or simply living amongst the night people. As a class, we watched one another fall into those periods and come out of them intact. We had become a close-knit group, and it was hard to watch a friend slowly deteriorate physically and emotionally over the course of their night float weeks. We became used to seeing growing bags under our eyes and hair that wasn't recently groomed. In class, we adopted an unspoken leniency when someone couldn't quite get themselves to participate because they had been worked so hard the prior week and their sleep cycles hadn't yet recovered. It was a kind of tacit camaraderie I wouldn't wish upon anyone. There is a longstanding balance that residency training programs must strike between continuity of care against the cost that inhumanely long shifts incur in the form of sleep deprivation. There are many irreplaceable lessons to be learned from the work of treating patients from the time they get admitted to the time they're discharged, but the process takes a toll. Residents become exhausted and depressed without adequate rest, and they can eventually succumb to a form of moral injury if they do not feel the positive reinforcement that comes too sporadically from empathic supervisors and appreciative patients. Even with duty-hour restrictions being put into place that require more time off away from the hospital, many residents find it hard to balance their performance and their morale.

After exiting my own night float rotation, I did eventually recalibrate to the world and started sleeping at night again. I decided to reach out to Ashley by text. She was responsive but seemed distant. Well, she *was* distant, being in Arizona, but her responses weren't as warm as I was expecting. Pretty quickly, I got in my own head about it and decided that I would stop initiating texts for a couple of days to measure her interest level. This kind of secret test made me feel like the worst kind of crushing adolescent. My more mature brain told me to pick up the phone and call her, but online dating seemed to reduce me to the mind-set of a teenager. A day went by and I received nothing from her end. I began to fill in the blanks of my neuroses with all of the worst possibilities. *She had met someone else. She'd decided she didn't like doctors. She'd decided she didn't like me.*

On the second day, I was having coffee with Rachel and Miranda. I shared my insecurities.

"One day of silence is not so bad," Miranda said supportively.

"We're on day two now," I said.

"One day or two, whatever. You're a creeper no matter what for going out with a college girl," Rachel said.

"She's very mature."

"Then why does she like you?" she retorted with a smile.

Defeated, I took a sip of my coffee.

"What do you guys have going on for New Year's Eve?" I asked.

"I'm visiting my brother in Chicago," Miranda said.

"I may go out with my Medicine friends in Cambridge," Rachel said, noncommittal.

"I've got nothing. Will you let me know if you end up somewhere I can join?" I asked.

I couldn't think of anything more symbolically pathetic than spending New Year's Eve alone because I couldn't find anyone to spend it with.

"Okay, but I really don't know what the night's going to be like," she replied. "We might end up anywhere."

Her response made me feel like she didn't want me there, which stung, but I told myself to stop projecting my own insecurities onto her words. When New Year's Eve arrived, it had been days since I texted with Ashley, and I was desperate for a quick dopamine hit. She hadn't yet texted me since I began the secret test. I decided to text Rachel early in the evening to take my mind off it.

How's it looking?

No response.

Any updates?

An hour later she replied.

Not sure. Will text you later.

By 10 p.m. I was feeling worthless. I ran through my phone's contact list, but everyone I felt close to was either in New York, or married with kids, or — as in Rachel's case — ignoring my texts. I sank into the couch and stared at the wall. I couldn't believe that six months into my residency, I wasn't capable of securing an honest-to-goodness New Year's Eve date. My aspirations were modest. I wasn't even looking for someone to kiss at midnight. I just didn't want to feel so alone, and I was furious that Rachel was abandoning me even though I knew that she never said she would be my ticket to a good night.

I texted Ashley down in Arizona.

Happy New Year's Eve!

Why does one exclamation point seem so cheesy? I wondered. It suggested the naïve enthusiasm of a small child. Should I have used two exclamation points? No, that would have been too much. A period? No, periods in text messages are sociopathic. There's no winning with texts, I decided. I should have called her, but what if she was out with her friends, or worse, a guy? The relative patheticness of my New Year's with no plans would have made me want to shrink into nothing and never reappear.

I watched as the clock inched toward midnight, but Rachel never did text.

At 11:45 I wandered down from my apartment and found a bar nearby. I ordered a drink and sipped it quietly as the ball dropped on the big screen behind me. I watched as couples around the room celebrated and locked lips. I finished the drink and walked back home feeling like the biggest loser on the planet. Worse than not having any plans, I had gone out at a quarter to midnight solely for the purpose of telling anyone interested the next day that I had not stayed in.

As I got into bed, I received a text from Ashley.

Happy New Year, friend! it read.

Friend was the nail in the coffin. I wanted to crawl beneath the covers and never come out.

The next day Rachel told me that her Medicine friends ended up at a Medicine party and she didn't really feel comfortable inviting someone to someone else's house. I wondered if she kissed anyone at midnight, but I didn't ask.

"You've gotten pretty close with your friends on Medicine?" I asked.

"Yeah. Sometimes it's nice just to speak with people who aren't psychiatrists for a change," she replied. "Sometimes they even have normal, everyday conversations like regular people."

"We do that," I said, objecting.

"Not as much as you think. You'll see when you're on Medicine."

The truth was I was already dreaming about my upcoming change of pace on my Medicine rotations. I was terrified to masquerade as an internal medicine doctor, as all resident psychiatrists must do in their first year, but I couldn't wait to start the next chapter and get away from the feeling of utter loneliness that had begun to eat at me in Psychiatry.

I regretted that I wouldn't be able to watch as Deborah, hopefully, rebounded with the help of ECT. I wouldn't be able to see how Jane managed, but maybe that was for the best. I wasn't sure I could take watching her fade away. The pain of being so helpless to stop it seemed to be too much to bear. As a medical student, I had only really been exposed to the positive parts of the field. I never had the opportunity during brief rotations to see people longitudinally get better or worse over time. I had never experienced the sense of powerlessness I was having midway through my first year.

I was assigned to do my Medicine rotations at a community hospital twenty minutes south of Boston. It had a good reputation for training us in the "bread-and-butter medicine" problems we would need to know as psychiatrists — things like urinary tract infections and pneumonias that we would need to be able to recognize as psychiatrists because of the mental status changes they sometimes bring. Anything too medically complicated at this hospital was often bounced back to one of the major medical centers in the city.

I was surrounded by other doctors of all stripes who were doing a preliminary intern year in anticipation of careers in fields such as radiology, ophthalmology, anesthesiology, and, of course, internal medicine. There were stunningly smart young doctors who had purposefully chosen this clinical program for its reputation as a kind of breezy medicine experience, though within their chosen specialties they would go on to do great things at the most prestigious hospitals in the country. They had also already been practicing medicine for the last six months, while I was toiling away in the very different world of psychiatry, so their ability to simply get stuff done put mine to shame. I hardly knew where the prescription pads were located, and each of them seemed to have memorized the protocols involved in admitting, treating, and discharging patients with pneumonia, congestive heart failure, cellulitis, and all of the other super-common admissions to general medicine. It took me a full month just to find my footing in the computer system so I wasn't lagging behind everyone and slowing the team down.

Generally, teams comprised a senior attending, a single senior resident, and multiple interns. The entire operation felt like what I had always imagined military culture to be like. Interns reported to the resident and took orders directly from them. The resident reported to the attending, where the buck generally stopped, though behind closed doors they were answering to department heads and medical administrators. The entire dynamic was saturated with hidden angst and opportunities for error. The interns did not make many decisions, but if they didn't catch a careless error, a patient's entire treatment plan could be derailed. I spent so many hours putting in orders and writing up notes, tasks that had almost nothing to do with actual medical skill, and yet it was the most important scut work I'd ever done because it allowed for sick patients to engage in a system that would get them better.

Through my rotations, I became close with many of my Medicine co-interns. There was Herberto, the knowledgeable and kindhearted doctor from Mexico, and Helen, an eccentric woman from Montana whose exceptional warmth was matched by the disarming and peculiar way she interacted with the world. In her apartment, for example, she kept a life-size seven-foot cardboard cutout of NBA star Dirk Nowitzki in the main entryway. On the medical floors, there were intercom messages known as code blues for medical emergencies and code purples for psychiatric emergencies, but Helen invented the term code brown for an emergency involving a patient's bowel movements — on the Medicine wards, there were plenty of those.

Then there was Jack. I met him on my first day. He was tall with a boyish face and exceptional posture. He would pair kindly advice with gentle ribbing, which made me feel like we were buddies right away. He struck me as a genuinely good guy and a great person for me to ask questions to along the way. Even though he had his own slog of never-ending tasks to complete each day, he never admonished me for my incessant questions. *Where is the pulmonary embolism order set? Is the CT angiogram done with or without contrast? What is the creatinine cutoff for receiving contrast? Why won't the computer let me log in? What am I doing here? Who am I?*

Jack was tremendously kind to me, particularly early on when I didn't know what I was doing. I realized quickly that most interns are just faking it until they figure it all out, but the problem I faced was that my colleagues had done that six months earlier, which made me look inept to my senior residents and attendings. The upside was that there always seemed to be people like Jack around to guide me and seniors there to catch my mistakes, which just left the embarrassment of looking bad as my biggest problem. With no intention of ever setting foot in that hospital again after completing my Medicine requirements, I discovered a certain solace in being the slowest member of the team. I recognized that less was probably expected of me as a Psychiatry resident, and I could use that prejudice to my advantage and focus more on doing what needed to be done regardless of the pressure to perform for others. It was a freedom from performance anxiety I had never felt before.

I did rotations on the floors, which meant general-medicine

inpatient wards, followed by the intensive care unit, which featured overnight on-call every fourth night, and finally in the cardiac care unit, which required interns to be on-site for thirty straight hours every third day. I learned that there can be psychiatry, to a greater or lesser extent, in every patient encounter. All of us walk around with all kinds of anxieties and neuroses on our best days, and I came to realize that when someone is hospitalized for a medical condition those traits became magnified. Some patients admitted to Medicine, though, really belonged on Psychiatry because their physical symptoms were actually the result of psychological distress — patients with seizures that showed no electrical activity in the brain, for example. Patients didn't know I was a psychiatrist in training, but in the care of these patients, my team would inevitably look to me as a kind of psych whisperer.

"This is a great case for Adam," the senior resident would say.

I had looked to Medicine as a respite from the world of psychiatry, but the military order of things meant that I took what assignments were given to me.

Many of these patients responded quite poorly to my telling them that a diagnosis might be psychogenic. Some screamed at me, and others quietly sobbed about there not being an easy physical fix for what ailed them. I had become well equipped to manage those reactions with empathic listening.

By the time I rotated through the intensive care unit, I had learned the lay of the medical land. These were physically exhausting shifts. The year after I completed my first year of residency, tighter duty-hour restrictions were put into place by the accrediting body that governs all post–medical school training. These rules limited shifts to twenty-four hours with built-in days off for recovery, but they came a year too late for me to be affected while on my Medicine rotations. My experience in that first year was limited to thirty hours of consecutive work and no more than eighty hours of work per week on average across several weeks. In any particular week I could exceed that total.

The cardiac rotation took full advantage of this and was an experience I had never had before, where I truly felt like I lived in the hospital and the world outside was just a faint, distant backdrop. No period of my life ever passed so quickly, but at the end of it I felt completely spent. It amazed me to learn how much I was capable of after removing the stressors of the outside world by living almost entirely in the hospital. I worked more in that month than at any other time in my life, and while I was proud of what I accomplished alongside my team — the lives we saved and the suffering we kept at bay — I never wanted to do anything like that again.

I had two more weeks of 7 p.m.–7 a.m. night float, and then I was to have my first vacation in months. As a kind of silent act of rebellion, I decided to grow a 1970s-style mustache for that last stretch before vacation. I figured that if I was going to be confined to a life with only the night people, who chose to work in the dark and sleep in the light, I was going to have some fun with it. By growing a mustache, I felt like I was sending the message of how emotionally done I was with that rotation, and it made me feel ever so slightly better.

With my new appearance, I seemed to cause a double take and silly comment every time I walked into a room. Professional men my age simply did not wear 1970s-style facial hair. I was feeling fine about it until I received a page calling me to the ICU. Howard James had died.

Now, I had never met Mr. James. I didn't know Mr. James from Adam, but as the intern on call overnight, I had the duty of pronouncing him dead and then telling his family, who sat together in silence in the waiting area adjacent to the unit.

I marched into Mr. James's room and casually performed the type of assessment that just several months earlier had flummoxed me. He had no "doll's eyes" reflexes. His pupils were fixed and dilated. His blood was beginning to pool in his body, and his muscles were stiff with rigor mortis. There was no question that Mr. James was dead.

"Time of death, four twenty-nine a.m.," I announced to no one.

Walking slowly to the waiting area, where I saw his two daughters and wife huddled on the sofa in tears, I remembered I had that stupid mustache on my face. The family looked up at me desperate for information. I thought about running back out and grabbing a surgical mask or covering my face with my hand, but I couldn't leave them in any greater suspense than they already were, and I was sure they couldn't possibly care about my facial hair in a moment like this.

"I'm Dr. Stern, the intern on call. Howard has passed away. There was nothing more that could be done. I'm so sorry for your loss."

For a moment everyone in the room was completely still. Only the ticking clock made noise. The fluorescence of the overhead lighting beamed down like an interrogation lamp, and my face became flushed. Then Mrs. Jones put her arms around her daughters, one on each side, and simply said, "Thank you."

When I arrived home the next morning, I shaved my face and booked a trip down to New York to be with my family.

13

Stern and Sons

I remembered the talk David and I had as we sat in the old Jeep our father had given him as a hand-me-down a few years earlier. The year was 2002, and we were waiting for the gas tank to fill before driving home for winter break together. He was a senior in college, and I was a freshman. I had followed in his footsteps to Brown that fall.

"I want to be pre-med, but it's too hard. I don't think I can do it," I said.

"Too hard?" he asked, incredulous. "Everything good in life is going to be hard. You think any of those other kids in your organic chemistry class are smarter than you?"

"Some are," I replied.

"They're just working harder. If you spent all of your time studying, you'd be getting As. We both would."

David and I look a bit alike, but our personalities are very different. He's always been more talkative and outgoing. His optimism is never-ending, while I've always had more difficulty believing that things would go right for me.

One trait we share is that we had grown up admiring our father for being a doctor. From early in childhood we could tell how respected he was within the family, always the one who relatives approached with concerns. We saw the appreciation that his patients had for him in funny little ways, like the small holiday gifts they sent each year, and we grew up with a shared desire to earn that same respect someday, to work in a field that took advantage of our innate skills while giving back to society and funding a comfortable lifestyle. In the end, we both learned that having a natural proclivity for math and science was just the very first step of many in becoming a doctor. Most of the path from undergraduate to medical doctor requires an infinite resource of grit and dedication that both David and I had to humbly cultivate over years before we fully stepped into the identity of being a doctor.

Our father, for his part, largely discouraged us from pursuing medicine because he felt the field had become unrewarding in the era of managed care and diminishing autonomy, but he was wise enough to know that our career path was our own to choose. Our parents both supported us along the way, emotionally and also financially. We didn't even have to take out student loans, which afforded us the opportunity to pursue our dreams wherever they may have led us. Not everyone has this luxury, and truly, I don't really know if I could have made it all the way through those challenging years in medical school while worrying that I might fail with no safety net.

Coming home again as interns on break, we fell into old routines, like bingeing on snack food from the kitchen cabinets that had been well stocked by our mother in preparation for our arrival. We also discovered new habits, like the medical one-upmanship we couldn't seem to stop.

"And then the resident was paged away and left *me* to do the rest of the central line by myself."

"Yes, but have you ever done one on an acutely manic patient who was trying to strangle you?"

"No, and neither have you!"

We were adding a ridiculous new spin to our age-old sibling rivalry.

On and on like this it went, until eventually I think we both realized that we didn't really want to play that game for the rest of our lives. I realized that even if I had been enjoying the competition, it would be impossible to compete with David as I returned to psychiatry and he continued along the path of internal medicine. Over time, we backed away from that brotherly competition, and I think we are both happier for it. At times I envy the bond my brother and father share as medical doctors. They can celebrate victories and debate dilemmas as father and son and as senior and junior colleagues. Psychiatry can be a lonelier endeavor.

I met with one of my best friends from high school, Jillian. It was unseasonably warm, and so we decided to get together by the duck pond near her house. It felt surprisingly good just to be in the open air.

"You look exhausted," she said, pausing for a response.

She would have been a very shrewd psychiatrist if she had been inclined to pursue it because she seemed to have a gift for seeing the truth in people and opening it up for analysis.

We hugged, and it felt so good to touch a human body without the clinical distance of a physical exam.

"So how *are* you? Tell me everything."

I could only sigh and look away.

"What is it? What's the matter?"

"I just don't know if I can do this."

"Do what?"

"Medicine, psychiatry. All of it. It's all so exhausting and isolating. I'm surrounded by people all hours of the day, and somehow I have never felt lonelier. I don't know. Maybe I should do something else."

"Something else? But you love it. You've loved it for years. It's who you are."

She stopped me in my tracks, and I looked at her. She had known me so well and for so long that I suspected she was right, but why didn't I feel like being a doctor was who I was?

As though reading my mind, she continued, "You need to finish up medicine and get back to the world where you can see patients as people. I think you'll start to feel more like yourself again when you're back."

"I hope you're right," I said.

"I am," she said, smiling.

Back at home the next day, my father approached me.

"I need to curbside consult you on a case."

"You do?"

He nodded.

"Middle-aged woman. Depressed. Had previously done well on fluoxetine. Says it stopped working. I'm thinking of trying her on duloxetine next. Is that a good idea?"

I asked more about the woman and what predisposed her to depression. He told me about the kinds of symptoms she had been having and what other medications she was on. We discussed potential side effects and even the risk of drug-drug interactions from her cardiac regimen.

"Still, it seems like it's worth a shot to me."

"How do you dose that drug?"

"I usually start out at twenty milligrams and go up to twenty milligrams twice a day before titrating further up to thirty milligrams in the morning and night or even higher."

"Very good. It's nice to have a psychiatrist in the network."

14

A Face Worth Licking

As my rotations on Medicine were winding down, I looked forward to reconnecting with my Psychiatry classmates. I also hoped to stay in touch with the gang from Medicine, whom I had come to admire. Despite our crammed work schedules, I had been able to remain connected with Jack, Herberto, and Helen, getting together every few weeks. We often met up at a bar downtown or a club with music. One night out, I got an unexpected text from Rachel. She had a college friend in town and was looking to show her a good time.

I'm just out with my Medicine friends, I responded.

The script had flipped since New Year's Eve.

I waited a minute and then added, You can join if you want.

She agreed and met us out at a club where they were playing loud electronica that was hurting my ears. I pretended to enjoy it.

"This is my friend Caitlin. We went to school together."

"Good to meet you, Caitlin. I love this place!"

"It's a little loud," Rachel responded.

Helen spotted me across the room, and as if feeling territorial she came over and put her arm around me.

"Take our picture," she said, handing her phone over to Rachel.

Looking slightly taken aback, Rachel took the phone and began to frame a shot. Just as she put her finger on the button, Helen turned toward me and licked the side of my face, tongue out several inches all along my cheek. It was truly vile, but oddly, I appreciated it. Maybe it would convey to Rachel that I was desirable, enough to lick anyway.

Helen grabbed her phone back and scampered off back to Jack and Herberto.

"Interesting friends you have," Rachel noted.

Later, I reconnected with Helen.

"What was that earlier?"

"That's the girl you're always talking about, right?"

I nodded.

"Had to show her that she can't wait around too long."

"And you showed her that by licking my face?"

"Yeah. How else would I?"

I shrugged.

"Fair enough."

Two days later Rachel told me that we were going to Mexico and we'd have to share a room.

"Come again?"

"Look, all of the Psychiatry residents have off at the same time in the last week of June. It's a quirk of the one-year contracts we signed because of our intern orientation," she said.

"So?"

"So, it's probably the last time in our entire lives we'll all be off from work at the same time. We should celebrate."

"By going to Mexico?"

"Yes. And I've already got everything planned. We're staying at an all-inclusive resort where we can just go to the beach all day and eat and drink as much as we want all night. Miranda and Gwen will both come from a Medicine trip, so they're staying together. Erin will probably bring her husband, Bobby. And no one else is going to want to come, which leaves you and me."

"Sounds fine," I said, wondering where her mind was exactly. This was not so much an invitation as a statement of how it would be, which was exactly Rachel's style. But was her arranging to stay with me a sign that she was interested or rather that we were *so* platonic that I was no risk?

"Definitely the former," Helen said when I posed the question to my Medicine gang.

"Agreed. On your first night there, you've just got to make your move," Jack added.

"And risk being rejected and spending the rest of the week awkwardly avoiding my roommate? I think I would die."

"Worth the risk," Jack said.

"You will not be rejected," Helen added. "I saw how she was looking at you at the club."

"It's a good thing you guys are better at practicing medicine than you are at doling out the relationship advice," I concluded.

That night I shared the dilemma with Magoo, my impartial and increasingly uninterested guinea pig, before going to bed, but she was no help whatsoever.

My very last shift on Medicine was ending, and I couldn't have been more excited to move on. The intense schedule and the just-following-orders lifestyle had worn me down. If I never entered another medical order set, I could die a happy man. I also looked back a bit more fondly on my days on Psychiatry, when there was time to use the restroom.

As fate would have it, though, my last shift was an intense assembly line of challenging patient encounters with medically complex individuals whom less sensitive doctors would refer to as "train wrecks." One patient after another seemed to be failing. One lost consciousness due to low blood pressure and was transferred to the intensive care unit. Another had the opposite problem and developed hypertensive crisis, landing him in the same place. A third literally crashed onto the floor, hitting his head, which earned him a full neurologic work-up. By the time Jack arrived as my relief and we sat down to do the patient sign-out — a tag-you're-it kind of handoff between doctors that occurs when shifts change — I was at my wits' end. Our senior resident casually walked over and, without making eye contact, let us know that the gentleman in bed 9 needed to be started on blood thinners.

"So you've got to do a rectal exam," he said to me.

"What! Why?"

I had done well enough as a psychiatrist in disguise for much of my Medicine experience, but every once in a while my cover got blown.

"He has a history of colitis," he replied while taking a bite out of his apple.

"So?"

"You have to make sure he's not bleeding before we thin his blood."

"You want me to walk up to a patient I've never even met, introduce myself, and tell him I have to stick my finger up his butt just in case he might be bleeding from his rectum? And then you want me to actually do it?"

"Yes, I do."

With that, the resident strolled off, taking another loud bite as he turned around the corner.

I made tentative eye contact with Jack, to whom I was signing out. I knew that technically the job fell to me because the order came in before we had finished the sign-out process. My mind wandered off to the blue-green waters of the beaches in Mexico and then to the shared hotel room with Rachel that awaited me after this final soul-crushing task that would conclude my training in general medicine.

"I got it," Jack said compassionately.

"Oh, Jack, I could kiss you."

"Save it for Rachel, buddy. Let's get you the hell out of here already. What are you even doing here? You're a psychiatrist."

"Yes, I am."

After placing a hand on his shoulder, I finished the sign-out and hung my white coat on the wall. Psychiatry, Mexico, and Rachel awaited.

15

Tongue-Tied Cherry Stems South of the Border

The Jacuzzi tub was clearly meant for two.

"This is nice," Rachel said. "Do you think they ever clean it though? I don't like the thought of what's growing in those jets."

"I bet they do," I said, imagining a scenario where somehow we'd end up in the tub together. "They must."

The all-inclusive resort Rachel had booked us into seemed nice enough. It featured several three-story buildings scattered around a well-manicured landscape with pools, palm trees, and the beach at one end. Our room was entirely nondescript, but at least it wasn't a shabby mess, which was a distinct possibility, since we had booked through an *extremely* discount online travel agency.

"I want that bed," Rachel said, moving toward the queen bed closer to the window.

Okay, so she wanted separate beds. I felt like a detective looking for clues about Rachel's interest in me and coming up empty. In our first six hours traveling together, I hadn't found a shred of evidence to support the position that we might be there as more than friends.

We changed into casual evening wear and walked through the outdoor complex past seductively lit pools to the restaurant Erin had chosen, Japanese-themed serving mostly Mexican food.

We walked in and Erin was already at the table. She smiled up at us, beaming.

"It's unlimited drinks, but they're all watered down, so you have to order a few at a time and mix them together."

"Where's Bobby?" I asked.

"Decided he didn't want to come at the last minute. Something about, if he wanted to hang out with a bunch of shrinks, he would check into 4 South. I really don't know, but I was not going to miss it. I am here!"

"And we're glad for that. It's not a bad place," I commented.

"Beats 4 South," Erin replied.

I wondered if it also beat time at home with Bobby, who had continued to seem universally miserable for much of the year.

"Both 4 South and this place do have unlimited snacks," Rachel added.

Miranda and Gwen arrived. Having traveled together for several days already, on a trip with their Medicine friends, they were clearly getting on each other's nerves. They were at each other's throats right off the bat, like an old married couple.

"We've discovered that we don't travel very well together," Gwen said.

"Alcohol helps," Miranda added.

"No, it definitely does not," Gwen replied.

"That just means you haven't had enough," Miranda retorted, signaling to the waiter.

When our drinks arrived, we all raised a glass.

"To one single glorious week away from 4 South," Gwen said.

"Away from Dr. Song," Rachel added.

"Oh, I miss him already," Erin said.

"Away from prescription pads and order sets," I said.

"And away from psychotics, depressives, and borderlines," Miranda said.

"Present company excluded of course," said Gwen.

"Ha ha," Miranda replied sarcastically. "Has anyone seen the ocean yet?"

"Miranda, we literally walked right past it on our way here," Gwen said with a hand on her forehead.

"We did?"

A waiter arrived with a second round of drinks.

"Just assumed you were ready for another," he said.

Then another came and another. Then we walked from that restaurant down a single path to a bar where we each drank more. It was the exhale at the end of a long year, and it felt great.

We settled with our bright blue drinks in oversize glasses onto a tented patio that had futon mattresses laid out on hardwood floors. Miranda insisted on showing us that she could tie a cherry stem into a knot with her tongue. Not to be outdone, Rachel matched her and raised her by a cherry stem. We told war stories like soldiers after the war had ended and cackled into the night, incredulous that we had survived the first year. Then, in the middle of a conversation, we noticed that Miranda had fallen asleep sitting up. Her head was tilted back and to the left, and for reasons that were clear at the time but have since left me, I took photos of Rachel and Gwen pretending to be vampires, baring their teeth toward Miranda's exposed jugular. It was shortly after seeing the expression on one of the waiters' faces that we all decided it was time to head back to the rooms.

On our way, we walked alongside a large pool that was immaculately lit in blue and purple from beneath the surface.

"We have to go in," Gwen said almost stoically.

"I don't think we're supposed to do that," Erin replied.

"Says who?" Gwen asked.

"Well, those guys, for one," she replied, motioning to two hotel security team members.

"They don't seem too concerned," Rachel said. "I'm not going first. Adam, you go."

As I couldn't be seen to be timid in Rachel's eyes, I took my shirt off and tossed it onto a nearby lounge chair.

"Let's do this," I said as I waded in, still wearing my blue jeans.

"You're wearing your jeans? Lame," Rachel announced.

The women then took turns taking their clothes off down to their underwear and bras.

"Really? I am not doing this," Erin said.

"It's no more revealing than our bathing suits, Erin."

"Whatever, those security guys are staring at us. You go ahead though. I'll try to scrounge up enough money to bail you guys out."

Rachel, Miranda, and Gwen joined me, and within ten seconds from the time they jumped in, one of the security guards blew his whistle.

"Hey, you can't swim now. Pool's closed," he shouted.

"Did you really just watch us disrobe all the way just to blow your whistle after we jumped in?" Gwen asked.

The guard just shrugged.

"Out now!" he shouted.

Dripping wet and shivering, we ran back to our rooms.

When Rachel and I entered our room, teeth still chattering, we walked right past the enormous bathtub.

"A hot soak would be nice," I said, hinting.

"For me, yeah," she replied, throwing me a towel to dry myself off before closing the bathroom door in my face.

In residency, there's a concept known as a black cloud. It refers to having bad luck when you're on call. Residents with self-declared black clouds always had the roughest, toughest, busiest calls while others skated by with a full night's sleep. The next five days, we literally all had a black cloud over us. There were almost nonstop torrential downpours. It rained on the beach. It rained at the pool. It even rained on us at brunch, leaking through the straw rooftop above us. We made the most of it with simulated good cheer and a fair amount of alcohol, but it was rough luck. Sometimes we played silly indoor games meant for junior high kids just to give us something to do while we continued the unwinding process, and that seemed helpful. A certain level of regression from our ultra-serious professional roles was a must just to recharge. After all, when we got back stateside, we faced what our senior residents had told us would be the hardest year of training.

By the last night of our trip, I was psyching myself up to make some kind of romantic move with Rachel. I had no idea what it would be. I've never been particularly confident with that kind of thing, but the opportunity seemed too good to pass up, and I remembered the advice from my Medicine crew.

"Good trip," I said, looking at her under the stars as we headed back to our room from the beach.

She kept walking in silence. After so much precipitation, the sky seemed eerily clear and you could even make out the faintest glow of the Milky Way galaxy overhead.

"Shame we can't do it every year," I said.

"Would have been better if it didn't rain every day," she said.

"It's beautiful tonight. Maybe we can still salvage the end of it. Want to stay out on the beach a bit longer?" I asked hopefully.

She shook her head.

"I'm pretty tired."

We were thousands of miles from home under a beautiful sky all by ourselves. It really seemed like it was then or never. I stopped in my tracks.

"What is it?"

Why couldn't I just do it? Kiss her. Take her hand. *Something*.

She stood several feet ahead of me.

"Nothing. Just couldn't find my key, but here it is."

I began walking toward her, and she put her arm around me.

"I need to ask you for some advice," she said.

"Yeah?"

"There's this guy I met on Medicine who I kind of like, but I have no idea how to do it. I'm like the worst with all that stuff."

It was a question that seemed destined to crush my heart. If I were being honest with myself, she hadn't given me any real signs all week that we were anything more than comfortable friends.

"You're the worst at courtship? I think I've got you beat," I said. "I don't know. I wouldn't take any advice from me."

Part 2

Year Two

16

Like the First Year, but More

By our second year, our entire class appreciated just how essential it was to spend time together as a group outside of the hospital. The trip to Mexico had really bonded all of us. Those of us who had gone on the trip had those shared memories, and the members of our group who hadn't joined us were doubly motivated to cultivate a life beyond 4 South. Except for the few of us on call at the hospital, we all joined together for the first monthly edition of Psych Cinema, for example. It was an optional, extracurricular perk of being in the residency. Faculty members took turns choosing a film they thought explored concepts in psychiatry and invited residents into their home to watch and discuss over food and drinks. When I think about the warmth of our program, it is the extracurriculars such as Psych Cinema that come to mind. We also enjoyed book groups, ugly sweater parties, and annual weekend retreats to a camp up in New Hampshire.

As the movie *E.T.* played in the background, my formerly frazzled senior resident, Rebecca, now exuded the relaxed confidence of a third-year resident. She gave me the skinny on year two of residency.

"You'll do fine. It's just like the first year, but *more.*"

"More what?"

We were given a look by the head of the department, Dr. Philip Brown. He was a very kind and knowledgeable man who was hosting that month's Psych Cinema. In that moment, though, the departmental leader who preferred to be called by his first name but who was most often referred to by his last was shushing us with a glance and raised eyebrows as a family member would under the same circumstances. I motioned for Rebecca to join me in the kitchen to freshen our drinks.

"More what?" I repeated.

"Oh, goodness. More *everything.* More patients each shift. More shifts. More sleep disturbance. More cafeteria food."

My shoulders slumped. She poured me another glass of wine.

"Is there anything good about being a second-year?"

"Oh, for sure. For starters, you finally get to do only psychiatry! You're finally a psychiatrist again! Yay!"

"Yay," I repeated flatly.

"And you finally get to feel like you know what you're doing, but actually it's much more than that. By the end of this year, you're going to feel like nothing can touch you. Anyone who walks into the ER naked and screaming, anyone who threatens to sue you for malpractice if you don't discharge them, anyone who tells you that you're the best psychiatrist one day and the worst the next, you can handle it *all.*"

"Hard to imagine."

"You'll get there. Trust me. Oh, and you'll definitely take on your first long-term psychotherapy case, which is fun if you're into that sort of thing."

I could not wait. I was quietly putting up with the early rotations on all of the inpatient services as a means to an end, a career as an outpatient psychiatrist. I wanted to treat patients who wanted to get better by voluntarily coming to see me instead of the people who were often committed against their will on 4 South. Working on a locked ward never felt quite right for me, but this year I would at least get to start with one therapy patient who might be with me for the next three years. That was longer than most of my romantic relationships.

Feeling rejected by Rachel without her even knowing it, I met Rebecca's eyes. Wouldn't it be nice to have a girlfriend, to love and be loved, I thought to myself. As if reading my mind, she interrupted the fantasy.

"You're still not going to have time to date anyone, by the way."

"Dang. May as well get back to *E.T.* I guess."

"Yes, let's.

I didn't like the sound of carrying six patients at a time on 4 South instead of four, and I certainly didn't like the idea of more overnight shifts in the emergency department while managing the entire hospital's psychiatric needs with only a brand-new intern by my side. I had minimal confidence that I could mentor anyone, let alone guide them in their first psychiatric call. Then I met the interns and realized how pitifully little they knew. Some couldn't pronounce the names of the drugs we were prescribing. It shocked me to realize that we must have been just as helpless a year earlier. It also made me think back to my own entry one year prior and appreciate for the first time just how far I'd already come. The new residents were green and eager, while our group of second-years was already tired and jaded, but at least I had gained a sense of what the job was and what was expected of me. Unfortunately, what was expected in the second year was an awful lot.

There may as well have been ominous music playing as I swiped back into 4 South to start my second year. I dreaded those first five steps onto the unit and past the black line taped to the floor designed to keep patients from eloping. I said hi to Reggie, who gave me no more than a casual nod in my direction.

"Dr. Stern. Welcome back. I trust Mexico treated you well?"

It was Song. How did he even know about Mexico?

"It's good to see you, Dr. Song."

"Yes, yes," he said, waving his hands wildly, "enough with the pleasantries. We have work to do."

Working with Song and Crystal again felt surprisingly refreshing. It had been months since I felt the rhythm of a psychiatric team at work. I was excited to ask how patients were doing and then really care about their answer at least as much as I cared about any vital signs or lab values, though those remained important on 4 South as well. In fact, while I had previously learned that there was psychiatry in almost every Medicine case, this time rotating on 4 South made me realize how much internal medicine manifested on a Psychiatry unit. In addition to the many medical illnesses that patients walked in with, many of our psychiatric medications can cause a variety of side effects, impacting blood pressure, heart rate, muscle tone, and blood sugar, among other traditionally medical data points.

Early in the new year, a patient knocked on the resident-room door with real urgency. I opened it to find that his neck was twisted more than ninety degrees to the left, and he looked like he was in agony.

"Need help," was all he could say.

For a moment, my mind went totally blank, and I'm sure I looked like a deer in headlights. After taking a breath, though, I recalled a known potential side effect of antipsychotic medications in which muscles torturously tighten up in the neck, shoulders, and elsewhere. I knew enough to diagnose the issue and treat it with a medication called benztropine.

"Let me get you some medicine to help," I said. "I'm sorry you're having this reaction. You'll feel better soon."

When the tightness in his neck resolved, I felt a strong sense of accomplishment even though I was the one who indirectly caused the problem with my medication. Dr. Song patted me on the back when I told him the story, and by the end of the shift the patient even came over to thank me.

I felt like I was just finding my groove when an email came in from Dr. Brown.

> *Effective immediately, Dr. Song has resigned his position. We thank him for his service and the care he has provided to our patients. We will be organizing appropriate coverage in the coming days.*

Um, what?

Our first Feelings class of the second year came at us like a tornado. Before we could even discuss the Song news, there were three totally unfamiliar faces in the room. With strangers sitting at the table, it felt like the ultimate safe space was no longer safe.

The first woman to introduce herself was sitting right next to

Nina, who had guided us so gracefully over the prior year even when sometimes, collectively, we could be a hot mess. Now, introducing herself as the new co-facilitator of our Feelings group, here was Jen, who in many ways seemed to be Nina's polar opposite. Whereas Nina was tall and earthy, Jen was quite small and immaculately attired. In her introductory speech to us she told us that she had heard such interesting things about us from Nina and how much she was looking forward to working with us, but the faces around the room reflected a shared sense of betrayal. Who was this woman? We were doing fine with Nina. We had confided in her, and she had led us down a path of self-discovery and professional growth. Why did we need Jen?

When Jen finished speaking, our attention turned to the man sitting across from me.

"Hi, I'm Drew. I'm passionate about the intersection of psychiatry and neurology. I finished my neurology residency last year and will be joining from here on out. Looking forward to learning from all of you."

We already knew about Drew. We had been hearing about Drew for months because his spot had been kept open in our intern year, necessitating additional overnight and weekend call shifts for each of us. Before we ever got to meet him, we felt resentful about this person. We were also silently irritated by this new guy who thought he should be both a neurologist and a psychiatrist—working with the yin and yang of the brain's medical needs. Psychiatrists work from the perspective of a bio-psycho-social model that takes into account patients' brains, minds, and lives. More often than not, psychiatrists view patients from the top down, starting with who they are as people and moving on to their symptoms, while neurologists may tend to view the brain from the bottom up, starting with the very building blocks of neurons that form networks that communicate through electrical and chemical signaling.

Drew didn't see the world or our patients in these limited ways. He had the logic and reasoning of a seasoned neurologist, but very quickly he began to show us that he belonged with us, too. He never viewed a person as the sum of underlying parts, and more often than not, his neurologic training helped him see the whole picture while the rest of us were still trying to piece it together. He was doing both residencies — a rare and frankly absurd concept given how many years such an approach takes — because he wanted to specialize in functional neurologic disorders, which is medical jargon for ailments that land between the fields of psychiatry and neurology. As I had seen on Medicine, for example, the patients who appear to have convulsions (with no underlying seizures and tremors) that come on only with family present and with no concurrent physiologic signs of illness are thought to have functional neurologic syndromes. These patients are often the orphans of modern medicine. Neurologists diagnose the issue as psychiatric, and psychiatrists aren't usually well trained in how to treat these conditions. Drew wanted to change all that. He wanted to be the person who bridges these two fields, and once the rest of our group realized that, it became a lot easier to understand him and forgive him for having given us each an extra call night in the first year. More than once by the end of residency, he helped my patients and helped me save face by filling me in on a neurologic interpretation of a patient's symptoms that I might have missed entirely.

Finally, it was time for the mysterious woman to my left to reveal herself. We knew why she was there. One of our classmates from first year had left the residency to join her husband in the Midwest, where he had found an exceedingly rare tenure-track job in his corner of academia. So even with Drew's arrival, we would have been down to fourteen if we hadn't searched for a replacement resident. The challenge in replacing a resident in the middle of a program is that the numbers of transferring residents are extremely small, and the process is not governed by any kind of match as our original fate had been. Instead, it was dependent on program directors such as Dr. Redding reaching out to one another and asking if anyone was losing a resident. Though it happens rarely, residents can transfer out for all kinds of reasons. Sometimes it's family that disrupts their training in one program. Other times, residents leave programs because they simply weren't a good fit—either on the resident side or the program side or both, something did not mesh. We had no idea what this woman's circumstances were, but she sure knew how to make an entrance.

"My name is Svetlana," she said with a notable accent. "I am originally from Russia, obviously, but I have been here for number of years and am glad to join your group."

With exquisite posture, Svetlana was shy of five feet tall. She wore stiletto heels that gave her another several inches, and her lips were perpetually glossed—or maybe they just appeared that way naturally. Her clothes conveyed that she was proud of her figure, and her attitude seemed to be that she was going to be who she was, and the world could just get used to it.

"I am tough. I was in army before medical school," she said.

I thought she brought a bit of flavor to the group. A class might not thrive with fifteen Svetlanas, but one amongst the other fourteen of us type-A, buttoned-down shrinks in training worked pretty well. I suspected that her tendency toward radical transparency might help keep us honest and down-to-earth ourselves.

There was a palpable unease in the room, though, with both of our new classmates, a new co-teacher, and the news about Song's resignation.

Svetlana dove right in by saying what was on everyone's mind.

"So, what is with this email about the attending on 4 South?" There was silence as Nina and Jen surveyed the room.

"Hm? There was an email?" Jen asked.

Nina and Jen were on the Harvard faculty but not formally within the residency staff, so they had not heard the news, and no one wanted to answer. We were all in shock. The awkwardness of the lingering silence began to fill the room.

"There was an email that an attending on 4 South resigned. Today I think. Was he no good?"

Svetlana was doing the work for the group. Her recent entrance into the group gave her the space to ask the question we needed to answer. Why was Dr. Song leaving?

"Dr. Song was the best teacher I've ever had," Erin said. "No one cares about his patients as much as he does, and no one works as well with the difficult cases. This whole situation stinks. Like, it really stinks. Something is not right. Was he fired? If so, we have to organize a protest because he's the only person around here that really cares about me — about us, I mean — as residents."

She had tears rolling down her cheeks. I wondered if she ever felt comfortable revealing this side of herself to her husband,

Bobby. With the time we had spent together in Mexico, I had seen a looser, more relaxed side of Erin than ever before. I knew that Bobby was miserable in Boston and wondered if there was room for her to exist fully with him as a cohesive unit beyond that misery.

"It sounds like you like this Dr. Song?" Svetlana said.

After a long pause, Erin sighed and looked up from her lap.

"Song and I had almost become friends over the last few months. I've been going through a hard time, which I don't want to get into right now, and it seemed like he was always there for me when I was having a rough day. Like with his patients, he always knew the right questions to ask and which ones to let pass. He cared about me even when it seemed like no one else did. Not my husband. Not my mother. No one else."

"We all care about you," I said.

"I know. It's just we're all busy with our own lives. I get it. Song always made time for me, and now he's just gone? Just like that? I had been getting the sense that he was unhappy here. I think he felt mistreated, and I know that some people in the department were not fans of his eccentricity. Like it matters that he doesn't wear a tie? Give me a break." She paused again to wipe tears from her cheeks and gather her composure. "We have to do something," she concluded.

"Do something?" I asked. "Like stage a protest?"

"I don't know what. Maybe we can write a letter from all of the residents in the program."

"Saying what exactly? We don't even know if he was fired or if he just resigned on his own terms."

"Song wouldn't quit on us."

"You don't know that," I replied. "We are like foot soldiers here. We have no idea what goes on at the higher levels in the department. Maybe there's more to the story."

Erin was being irrational, but I also felt like my resistance was coming off like an attack. I could sense that she was fuming at me, but we were at an impasse.

"We're heading back to 4 South after this. Let's go talk to him. There's got to be more to this story," she said.

17

Unwelcome Departures

Dr. Song was whistling, with the door open, when Erin and I arrived in his office.

"You caught me performing one of my favorite pieces from home," he said. "I take it you've heard the news. Sit down."

He continued taking books off the shelf and placing them gently into a cardboard box. He held each book for an extra second as he reviewed the spine. An almost imperceptible smile crept across his face every time.

"You have questions? Concerns? Tell me."

"You're . . ." Erin began but had to pause. "You're what? Leaving? Why?"

"This chapter has finished. Yes, I am leaving."

"I don't understand."

"Truthfully, it's not my place to discuss all of the details, and thankfully it is not your burden to know them. The important thing is to —"

"No. You do not get to tell me what the important thing is," Erin said sternly. "You *had* the privilege of telling me what was important when you worked here. When you were our mentor.

When you were my friend and told me what was going on. I guess those times have ended, so I'll just get back to work."

She began to stand, shaking her head as she looked at the floor.

"But that is just what the important thing is, Erin! As always, you are a step ahead."

"Getting back to work is the most important thing?"

"Let me tell something to my star pupil," he began. He turned toward me and said, "You are okay, too, by the way. You will both make excellent psychiatrists someday. You are okay. I am okay. We will continue to *be* okay, but those patients in there," he said, motioning in the general direction of the unit, "do not have the inherent strengths that we do. I am leaving them to you, in your care, and I need to know that you will continue to work for them."

"Of course," I said. "We are contractually obligated," I added dumbly.

"You must continue to work for them even when the universe says that you should not. When the administrators are pushing them out the door and the insurance company is putting up mountains to keep you from them, you must still work for them with even greater passion. That is how you will become the magnificent psychiatrists that I know you will be."

"How do you know we're going to be okay?" Erin asked.

"I have seen a lot of residents come through this unit. In many the fire has gone out long before they ever even stepped foot onto 4 South. Your fire is not out. It is at risk. It is flickering and needs oxygen. I see it in both of you. The system has a way of beating down even the best of us. You are in the thick of it now, but you have come this far, and I have seen enough to know that you will emerge at the end of your training stronger for it."

We sat through several seconds of uncomfortable silence as Song smiled at us from across the desk.

"You're really not going to tell us if you were fired?"

"Fired? Didn't you read the email, Dr. Stern? I resigned."

"Yeah, but I mean —"

"Dr. Stern. This is your home for the next three years. Even if I had dirt to shed, and rumor mills to stock, what good would it do for you? You deserve a lovely home, and this has to be it. It used to be mine. No longer."

Finally, the smile had faded from his face.

"What will you do?" Erin asked.

"I have a connection at a program in Texas or maybe one of the Carolinas? I forget, but it's a former resident, actually. Pay is surely double, maybe triple what we get here. I will be fine. I promise you that. I have heard that it is always better to be *from* Harvard than *at* Harvard."

There was another awkward pause. Song stood up and opened his arms for an embrace. I began to lean forward but then realized it was meant for Erin alone.

"I'm going to miss you," she said.

"Of course!" he replied.

I put out my hand and we shook with a degree of false stoicism that neither of us could pull off very well.

"Take good care of yourself," I said.

"You take care of yourself *and* them," he replied.

The smile had returned to his face.

Erin and I walked out of his office and back toward 4 South.

We swiped in and almost instantaneously became engulfed in the milieu of the unit. Song's announced departure followed immediately by the grinding work on 4 South made us feel like children who had just been told our parents were divorcing but we should go ahead and focus on our schoolwork.

There was plenty of work to be done, though. We were covering more patients than ever before, and everyone around us seemed to need our help. There were prescriptions to write and checklist items to tick off. More important, there were people behind the diagnoses who needed our help. Song had taught us how to listen and in our best moments to guide those people forward to a better place.

By the end of the busy afternoon, Erin and I sat in our back office typing out discharge summaries.

"Erin," I began.

"Hm?"

"You and Song were pretty close."

"Yeah."

"Too close, do you think?"

"Yeah, probably."

I nodded to her, agreeing with my own assertion.

"I have a tendency of doing that," she said.

"Doing what?"

"Idealizing guys like Song."

"Guys like him?"

"Men who tell me everything is going to be okay. I need that a lot — probably too much — and Bobby can only give so much. I lean so hard on him at home. None of you see that. All you see is this sullen guy, but you don't see me going home to him crying because I don't know if I can make it in this place. Then when I start to feel like Bobby's at his wits' end with me, I have to look elsewhere for that support. It's not the first time I overconnected with someone who was supposed to be my teacher."

"This has happened before?"

"I'm pretty much a nervous wreck most of the time, in case you haven't noticed. When someone makes it clear to me that they'll show me the way, and expresses interest in me, and praises me, I fall pretty hard for it. Not romantically. Not exactly anyway. It's deeper than that."

"Where does Bobby fit in?"

"Bobby does the best that he can, but sometimes I think my neuroses about being the best and doing the best are too much even for him."

"How so?"

"We've been together since high school. He's part of who I am, and it will always be that way. We've grown up together, but it still just feels like we're two kids trying to survive. We're just terrified making our way through the world together, and it's better to be together in that than alone in it, but it's not the same as when someone can show you the way."

"Plus, why is he so unhappy all the time?"

She snorted with inappropriate laughter.

"You're right! Why *is* he so miserable all the time? Here we are in Boston, this great city, in the prime of our lives. He doesn't have to work. No kids to take care of . . ."

She paused and frowned.

"What is it?"

"I was pregnant last month."

My eyebrows hit the ceiling.

"It was nothing we planned. It just happened, but then it un-happened."

"Erin, I'm so sorry."

"It's nothing. I mean it should be like nothing; we were so early in it all. We're not ready to have kids, too, so it's probably for the best, but ever since it happened, I can't get it out of my head. It's like I've always been on this path where in each moment I have no faith in myself doing the right thing at all, but I have to fake it. If I fake it long enough, maybe in the bigger picture I could eventually be this academic superstar. Getting into this program was just another step toward academic world domination to me. I was going to be a big shot with publications and extra letters after my name, and maybe I could be an expert that people would care about hearing from."

"That's exactly how I picture you. You're the smartest person here, and you're driven like no one I've ever seen," I said.

"Yeah, but that's just it. When this tiny, invisible fetus showed up — for just the few weeks we were together — I felt like actually I didn't need to fake it. I felt like the two of us were good enough just on our own. Somehow the need I felt to take care of this little being gave me a sense of who I really am, and it felt so good. I think maybe I need that more than any of this other stuff," she said, motioning to our surroundings.

"So do you think you'll try again?"

"I don't know. I never even told Bobby I was pregnant."

My eyes widened.

"I know it's awful. I didn't know how he would react. The two of us together as a unit can barely manage, and he already feels like he has to hold me up so I don't fall. How can we manage a baby? Of course, I was going to tell him eventually when I figured out how, but then I miscarried, and it was just over."

"I'm no expert, but I think these kinds of feelings tend to linger and it might be good to share that with your husband."

"I know. I'm going to talk to him, I just—"

There was a knock on the door. It was a checks person letting us know that the new admission had arrived from the emergency department.

"Who's taking it?" he asked.

My eyes met Erin's, and her words about being taken care of reverberated through my mind.

In unison, we both said, "I've got it."

We smiled and I stood up.

"I've got this one."

"Thanks, Adam."

Rachel: i woke up at 5am

Rachel: fail

me: yeah, noticed you playing Words with Friends at an ungodly hour

Rachel: i guess you can only go to bed at 9pm for so many days in a row

me: the good news is that if you wake up at 5AM tomorrow you can go to work and get all of your notes done before morning rounds.

me: and then you can spend all day doing anything you want

Rachel: yeah

Rachel: but yet, stuck at work

Rachel: i'm sick of being a doctor

me: not much you can do about that i don't think

Rachel: yeah

Rachel: i just looked at my loan balance

Rachel: ouch

me: at least hang in there til you can start moonlighting and being paid a market rate for being a doctor

Rachel: yeah

Rachel: i actually think with moonlighting i can prob get my own place and not have roommates for fellowship haha

me: there's nothing good in the movies

Rachel: did you see the girl with the dragon tattoo

me: no

Rachel: it is questionable whether i'd be able to sit through it

me: yeah, don't know

me: my parents liked it, but who knows what that means for you

Rachel: haha

Rachel: i liked the book

Rachel: was underwhelmed by the swedish version that was supposed to be “really good”

Rachel: we could see the adventures of tintin 3d

me: i don’t think so

Rachel: now i want popcorn

Rachel: do you want to go see the girl with the dragon tattoo at 1?

Rachel: i’m bored

me: okay

Rachel: Fenway theater

me: alright, i’m gonna go take care of some things. see ya then

18

Snoozers Lose

Charlie was a patient already in psychiatric treatment, but he was new to me, and I realized that neither of us wanted to be engaged in this evaluation. His choice of words had taken the decision out of our hands though, as he had expressed what's known as conditional suicidal ideation, saying he would end his own life *if* the surgery revealed cancer—and it had, which meant that when he came to, he was met by a one-to-one sitter and a pink paper denoting that he was not free to leave even "against medical advice." Before his operation, he had told one of the attending psychiatrists that he didn't think he had the stomach to die of cancer and would rather end it all while he was still able to. I assumed that the attending psychiatrist, who had known the patient for years, was home with her family, but she had left a note telling the surgical team that if the biopsied tissue revealed cancer, he should be emergently evaluated.

I was paged to come see Charlie because I was the second-year resident on call. Even though I had never met this man, and even though I had little formal training in doing psychiatric consults on the medical wards at that point, it fell upon me to both figure out if he was truly a danger to himself and see if I could possibly offer some kind of therapeutic encounter. I would settle for doing only the former if that's all I could muster. After the interview, I was supposed to present the case to a senior resident by phone and the attending on call at home. This was a delicate dance among psychiatry residents, their supervisors, and our patients. Even when we made decisions in the room with a patient, we often had to clear them with a higher authority, so we tried not to make any promises in real time lest we reveal ourselves to be the impotent underlings that we were. In this case, it also happened to be the middle of the night by the time he perked up enough to talk.

"And who the hell are you, now? First I've got this lady," he said, motioning to the one-to-one sitter, "and now some Doogie Howser, MD, comes in. What do you want, Doogie?"

"I'm Adam Stern. Dr. Stern."

"Okay, Adam Stern, Dr. Stern. Why are you here?"

"Your surgical team asked me to come talk with you."

"You're not part of my surgical team?"

I shook my head.

"Then what fucking team are you from?"

"Psychiatry."

"Oh, for fuck's sake. Get the fuck out of here."

"I can't do that. I've been asked to assess you."

"Assess me? Assess this," he said, grabbing his crotch.

"I know you've had a difficult day. I don't want to take up any more of your time or energy than necessary, but I do need to evaluate you."

"Why is that?"

"Well, before you went under, you told your psychiatrist—Dr. Glidden, I think—"

"You think? Yeah, it was Glidden."

"You told her that if it was cancer, you would end your own life."

A somber expression fell over his face, and immediately I knew that no one had told him yet. *For fuck's sake* indeed, I thought.

"It is. It's cancer? Damn. I'm not surprised. All the signs were there, I just . . ."

He began to sob into his hands. They were heavy, deep, almost visceral wails the likes of which I'd never heard before. I looked over to the one-to-one sitter, as though she would know what to do, but she had her face in a newspaper. I felt the urge to put a hand on his shoulder or even hold him in my arms, but something held me back. Psychiatry is strange about physical contact with patients, in many cases for good reason. Even when common sense and an empathic stance dictate that a hand be placed on a patient, residents are trained to be wary and think twice.

"I'm so sorry," I said quietly, almost whispering.

He continued bellowing without relief.

Finally, I overcame my programming and put a hand on his shoulder. Charlie was a big man in his sixties. He reminded me of my father, and it felt out of character but still right to be consoling him in this way. With the weight of my hand on his shoulder, he stopped crying, momentarily pulled back, and then leaned into it. He looked up at me.

"Thank you. I'm okay now."

I leaned back in my chair and waited for him to speak.

"Cancer. What a clusterfuck. My wife is going to kill me. She always told me to take better care of myself. Drink less. Exercise more. I just never really thought it would happen to me."

"Do you have any sense of the prognosis? I assume not, since no one has talked with you yet."

"Well, no, but also there aren't a lot of *good* liver cancers as I understand it."

"That's fair. Let me ask you something. What's the most important thing to you right now?"

"How do you mean?"

"Well, what we know is that you have liver cancer, and we do not know much else. That part is out of our control, though. I'd like to think with you about factors that may be under your control."

"Like what?" he asked.

"Living as long as possible with a high quality of life. Preventing suffering. Making sure that your family is taken care of. Those kinds of things."

I was really out of my depth, but it seemed like we were connecting. He talked about trips that he and his wife had put off for years that maybe they would finally go on together. He spoke about living long enough to see his son get married.

"It sounds like you have a lot to live for."

He nodded.

"That comment that you made to Dr. Glidden," I started.

"Forget it," he said.

"No, I can't forget it. My job is to take it seriously. You have good reason to feel that way, and I don't want to minimize that.

But I do need to know if you are safe to be on your own right now."

"Today I'm safe. I won't do anything yet."

I had been taught that this kind of contracting for safety is generally insufficient in a psychiatric risk assessment. He had a number of major risk factors for suicide in the context of his new diagnosis including that his age was over sixty, and he was a male. My gut told me he wasn't going to act on the comment he had made, and that was also supported by his marital status and his being future oriented. I confirmed there were no guns in the home, though a simple rope can be just as lethal. He had no family history of suicide, which is a powerful factor in many cases.

"Are you going to send me to the loony bin?"

"I don't think you belong there," I said. "What do you think?"

He shook his head.

I needed one more scrap, just one little thing that would put my decision to let him off suicide watch on more solid ground.

"I have to do some work here tomorrow morning, and I should finish by about noon. Can I come back and see you before I head home?"

"I'd like that."

It wasn't in the textbook, but his willingness to acknowledge that he would be around the next day and allow me to visit made me feel more secure.

"Good, I will see you then."

But I had forgotten one of the cardinal rules of overnight consults, which was to never promise anything before speaking to the attending. Tony Strand, our psychopharm teacher, was on that night. I presented the case and tried to convey all the points working in favor of lifting the suicide precautions.

"It's a tough call. I don't feel good about it," he said. "Let's keep him on overnight and let Glidden comment in the morning."

"I think that's really going to throw this guy for a loop. I'm not sure if he'll ever trust another psychiatrist again," I said.

"For tonight, better safe than sorry. Good night."

He hung up, and it was my job to return to Charlie's room and explain that his one-to-one sitter would be staying at least overnight.

"Who the fuck does that guy think he is?"

"It's my fault. I shouldn't have said anything until I knew for sure that we would be able to take down the precautions."

He scoffed.

"I'll still see you tomorrow, right?" I asked.

He nodded.

"Okay, try to get some rest between now and then."

I exited the room with my shoulders slumped and made my way down the hall to the elevators. It was almost 3 a.m., and remarkably, I was caught up with work. I walked to the on-call room and sat at the desk. I watched the clock tick for two minutes before making my way to the bed. It would be two more hours before the bagel place around the corner opened for the day's business. My eyes closed for what felt like just a long moment, and I looked up to see the clock read 5:05 a.m. I jumped up and slid on my shoes. Essential caffeine and glorious complex carbohydrates awaited me.

On my way to the bagel shop, I crossed paths with Slippery

Nick, the trickster who had duped me for twenty dollars on my first day of orientation a year earlier. I had passed him half a dozen times since then, and we always simply avoided eye contact. On that early morning, though, in my semi-delirious state, I purposely slowed down to greet him.

"Morning," I grumbled.

"It's gonna be a hot one," he replied without missing a beat.

"You want an iced coffee or something?"

"I'll take an Andrew Jackson if you got it," he said with a winking grin.

"I don't carry cash when I'm on call," I said. "Really, it's true."

He shrugged.

"Have a good one, Doc."

As I entered the shop, there was one person ahead of me in line. She was also in scrubs, and looked to be about my age. Our bleary eyes connected.

"I can't believe I wasn't first," I muttered playfully.

"Hold on."

The woman behind the counter handed her a steaming cup of coffee, and she took a sip.

"What did you say?" she asked.

She had the kind of face that, even when it looked tired, seemed to beam kindness out to the world.

"I've been waiting all night to be first in line but didn't get myself out of bed until 5:05."

"Snoozers lose," she replied with a shrug.

"I think the expression is, *If you snooze, you lose,*" I said.

"Are you mansplaining snoozing to me?"

"I would never. Well, certainly not at this hour. I'm Adam."

"Jessie. You just finishing a shift?"

I nodded.

"Psychiatry resident. How about you?"

"Just coming on actually."

"At five in the morning?"

"I like to pre-round before my pre-rounds before my rounds."

"Yuck. That's the worst thing I've ever heard," I said.

Rebecca had once told me that how residents act in a sleep-deprived post-call state is how they will be years from now when they sink into dementia. Some were grouchy. Others became dim-witted. I was disinhibited.

"Well, I only do inpatient work two weeks out of the year, so I like to make sure I don't kill anyone. It would be really tragic and embarrassing."

"Two weeks?"

"Yeah, two weeks on the palliative care service to pay my dues to the department, and the rest of the time I'm a pain specialist."

"Pain specialist? Like an attending?"

"Yeah, like an attending."

I had no idea I had been talking to an attending. She was so down-to-earth. She wasn't giving me orders to carry out or even trying to teach me anything. It just hadn't occurred to me that she might be a fully trained doctor. I felt like I had overstepped. It felt inappropriate to ask out an attending, even if she was in another field, but all the kindness in her face just kept beaming at me, so between the sleep deprivation and the oncoming delirium of the post-call morning, I decided to ask her out.

"I know this may sound forward, but I wonder if you might get a drink with me sometime."

"I am getting a drink with you. Right now."

She looked confused and my heart dropped.

"I'm just kidding, Adam. Let's do it."

She took my phone and typed her number into it.

"Give me a call when you've gotten some sleep, and we'll set something up."

The whole exchange made me feel like I was floating. I needed a boost after realizing that it wasn't going to happen with Rachel, and that Jessie was an attending already and was yet interested in me felt almost salacious. Since medical school, it had been drilled into my head that there was a pecking order that governed interactions amongst students, residents, and attendings. Even though we didn't really work together, the power differential felt very risqué and that much more tantalizing.

I coasted through my morning sign-out and rounded on a few patients up on 4 South before circling back to check in on Charlie. I rode up the elevator, still glowing from my encounter with Jessie, when a woman with very tired eyes introduced herself to me.

"I'm Charlie's wife. You're the shrink?"

"Oh, uh, yeah. I am the shrink I guess."

"You guess? Did you talk to my husband Charlie last night or not?"

"Y-yes, yeah, that was me," I stammered.

I was nervous that she was going to slap me for keeping her husband hostage on a psychiatric hold without any right.

"I could kiss you," she finally said.

"Excuse me?"

"Whatever you said to him really did something in that brain of his. He's been talking about offing himself if this was cancer for weeks, and now he's talking about wanting to fight this thing to the end."

It was a bit surprising. I couldn't think of anything I said to him that could have made that difference.

"I think maybe he just needed someone to be in it with him for a bit when things were really bad. I was glad I could do that for Charlie."

"Me too. Thank you."

We got off the elevator together and walked back to Charlie's room. He was sitting up in a chair, sipping coffee and reading a paper. I sat down with him, and he told me about the box score from the Red Sox game the night before.

"Can't believe I missed this game," he said, shaking his head.

He looked up at me.

"You look exhausted, kid. Are you free to go home yet or what?"

I nodded.

"Then get out of here already!"

I shook his hand and walked out the door. As was true for many consults, I left without knowing if I would ever see the patient again. I wondered how it would all turn out for him and hoped for the very best.

19

Tell Me What Brings You into the Office

While our class was beginning to find its footing in settings such as 4 South and the medical wards, we still felt like psychiatric impostors in part because we had no idea how to do talk therapy. Learning to do psychotherapy with patients is a little bit like ice skating. You can't really do it without putting yourself out there and falling on your ass a few times. Early in our second year, we had the benefit of being introduced to a woman in the department named Meg Mook. Dr. Mook actually had a PsyD degree, which meant that while she was not a psychiatrist, she had much more training and experience in doing talk therapy than many MDs. She had risen to a high position within the department of psychiatry despite not being a psychiatrist, which indicated to me that this lady must have been a real guru of psychology. As I got to know her over the course of the year, though, it became evident that even beyond her obvious competence, Meg had a powerful ability to connect with people, which probably enabled her to thrive in the challenging interpersonal laboratory of a prestigious department at Harvard. She was a tall woman with fiery red hair. She also simply exuded kindness and empathy that melted away the insecurities and neuroses that we each carried around with us. I've learned that when my technique and methods fail, reminding myself to begin with a foundation of Mook-inspired kindness and empathy usually succeeds.

Our first seminar with Dr. Mook focused on both broad strokes and fine points. We discussed theory and the fundamentals of different approaches in therapy — some therapies, such as psychodynamic and psychoanalytic work, were centered on uncovering unconscious conflicts. Others, such as cognitive behavioral therapy, were focused on more tangible concepts, such as changing behavior and thought patterns and eventually having a better relationship with one's own emotions. Some focused on interpersonal relationships, while others were largely concerned with the mind-body connection.

We also covered basic, practice-level questions we all had. What do you say to a patient as you walk back from the waiting room? How do you sit? How do you begin? What do you do if you run out of questions? The answers to most of these questions and many others were not singular but defined by a set of principles and applied on a case-by-case basis dependent upon what therapy is meant to be and what function it is meant to serve. If it is a path for patients to better understand their inner lives and, in turn, experience improved quality of life and less psychological distress, then it must be distinct from the common misconceptions many people have of it. It is not coaching, renting a friend, or no-matter-what-the-problem-is-it-started-with-your-mother.

It may be perfectly reasonable to chat with one patient in the hallway on the way to the office, while for another it could be taken as a betrayal of their privacy. One patient may be comfortable in a certain seating arrangement, while another would choose an entirely different setup. Sometimes silence can be a useful window into what lies beneath the veneer of a person's superficial experience.

Despite having learned better from Meg, I still managed to begin my first therapy session with an awkward speech about what I hoped therapy might be able to do for the patient. He frowned his way through it and nodded in unconvincing agreement until I was finished. I decided to never start like this again, and instead to simply ask the patient a version of *So tell me what brings you into the office.* I found out quickly that if I started that way and then just listened for a while, the patient would almost always provide much of the important clinical information without me having to ask another question. Unlike the fill-in-the-blank approach to history taking that medical students are used to, this approach allowed the patient to most efficiently communicate what was important to them, which is essential to the whole exercise.

After my patient tolerated my soliloquy, we sat in silence for several seconds. Meg had indicated that a good therapist becomes comfortable in silence for long periods of time, but at that moment it was hard to imagine ever feeling comfortable with any part of the experience. My own inner world was starting to crumble as I felt my cheeks flush and my heart race. I had staked my entire career on the idea that outpatient psychiatric practice was the thing that I wanted to do with my life. I was three minutes into my first session and already couldn't stand it.

It didn't help my feeling of being a fish out of water, flapping around clumsily, to know that the patient knew I was a novice at the job. I hadn't purposefully revealed that he was my first therapy patient, but he was aware that I was a junior resident who had no office of my own. That meant that I had to sign out someone else's unused office every time I would meet with him, which would often mean trying to become comfortable in a new environment every week. Residents were also generally assigned offices on the inner corridor of the department, with no windows or natural light. Whoever regularly worked in this particular office had done the best they could to improve the ambiance with mood lighting from lamps and even a painted accent wall, but it still screamed *low on the totem pole.* It was a far cry from the swanky psychiatrist offices I had seen portrayed on TV.

When eventually the silence was broken, it was by both of us simultaneously trying to clear the air.

"Go ahead," I said.

"The reason I'm here is that I think my wife hates me," Jim said.

He was an ordinary-looking man — normal weight, average height, just about my own age — with a full head of hair just beginning to thin in the front. It made me uncomfortable knowing that we had the same amount of life experience — by time anyway — and yet he was coming to me for help. Who was I to help him? I felt like an impostor.

"Tell me more about that," I said.

It was a good, open response that invited a patient to keep going. It was neutral and encouraging.

"Well, I'm not an easy guy to live with. I can be pretty tough on people because I expect a lot from myself and everyone around me. When someone lets me down — which seems to happen almost constantly — I can be rough. I mean, not physically. I'm not abusive or anything. Don't get that idea. I just get frustrated and make sure they know it. My wife — Shawna — she's wired differently than me. She's just so meek and timid and just tries to get through the day without pissing people off. It's infuriating to see how she goes through the world, but then when I try to pause and look at ourselves and how we live together, I realize that I'm not doing her any favors. I take advantage of her weakness without even realizing it. It seems like I'm always pissed off at her, even when she's hardly done anything wrong. So then she gets depressed, and I feel guilty, and that's when the shame really sinks in. That's when I start feeling like a real piece of shit. Like I'm always a piece of shit, but most of the time I just pretend I'm not, like I'm a good person. When Shawna and I get into it — which, again, is almost constantly — we both end up miserable and hating ourselves and each other. I don't know how I expect you to fix that, but I need help because it's not working out, and I don't know what to do."

I was far enough along in my academic studies to know that he was describing a classic archetype of narcissism. Beneath the gloating and bravado that can accompany a narcissistic personality, there is a pained person tortured by their own imperfection. While narcissists might show the world their puffed-up armor, in therapy they earnestly reveal themselves to feel like pieces of shit that the world seemingly revolves around.

My instinct as an inexperienced therapist was absolutely to pounce on this interpretation. I could give him the insight that he suffered from narcissism, change his sense of self in one swift motion, and he would be cured by the fifteen-minute mark. But Dr. Mook had been very clear that interpretations are best saved for a point at which you've already made an emotional connection with a patient. Were I to tell Jim during our first session that I thought I knew where his problems were coming from, it was likely he would be offended and either storm off or decide that I was a quack who had no idea what I was talking about, never to return. Rather, Meg would say, wait until a therapeutic alliance was formed and then the interpretation could serve as a narcissistic injury that could be repaired over time through the work of the therapy. I thought about what Meg had already taught me and decided to sit on my interpretation.

Every time Jim stopped speaking, I waited for several seconds and then asked him to *talk a bit more about that.* Through fifty minutes, I learned an incredible amount about his life and his relationship with Shawna. He was really cruising when I spotted the clock behind his head and realized we were over time already. I felt bad cutting him off, but I also knew that it was my job to be strict with the timing of our sessions. Defining the frame of our work with clear boundaries, I had been told, was incredibly important in long-term therapeutic work, particularly with patients who might struggle with respecting rules and norms out in the world. It took me two or three minutes to gather the courage to interrupt him.

"I'm sorry, Jim. I have to stop you there. That's all the time we have for this week."

"Wow. Okay. Well, thanks for listening to all of that. I feel like a weight is already starting to be lifted. I almost wish I could come back tomorrow to keep going."

"Let's shoot for the same time next week," I replied.

"Oh, okay. Yeah."

We shook hands and he left. I still had my note and billing

ticket to complete, but first I took a moment to stare at the ceiling and take in the milestone of having completed my first therapy session in one piece. The smile on my face faded as I realized that there were infinite potential sessions ahead of me — a career full of them — and sooner or later, patients would expect me to actually help them. I hoped that someday I would learn how.

Rachel: ugh drunk

me: do you need someone to hold your hair back?

Rachel: not that drunk

me: ha

Rachel: kind of headachy

me: fun while it lasted?

Rachel: it was really fun

Rachel: lots of good food and wine

me: mmmm

Rachel: ugh i think going to work this week is totally unnecessary

me: taking care of our patients is what attendings are paid to do, right?

Rachel: if i wasn't there, Fennington would kill my patients

me: hahhaha, intentionally or not?

Rachel: not

Rachel: he was like, oh we need to recheck absolute neutrophil count on weds

Rachel: i was like, uuhhh actually i put it in for mon, her ANC yesterday was 2

Rachel: he was like oooohhhhh

me: so if you don't go to work, people will die. there's your motivation

Rachel: lets start her on lithium

Rachel: i was like, well it's on her allergy list

Rachel: so maybe we should find out about that

Rachel: what did you do all weekend

me: nothing much

Rachel: how is ECT

me: really good, Macy lets me do everything. I've seen like dozens of cases already and I think I can pretty much do it on my own.

Rachel: that is mildly disturbing

Rachel: do you want to go to work for me tomorrow

me: yes

Rachel: hey, thanks

Rachel: i bet Fennington won't even notice

me: neither will Macy if you go for me

me: it's the perfect crime

Rachel: i need to go to bed and i don't want to

me: you do not need to

me: the night is YOUNG

Rachel: we should do another ski trip for new years eve weekend

me: why

Rachel: cuz i did it one year and it was fun

me: let's see how this one goes before planning next year's

me: do we need to go food shopping before hand or will we do it locally?

Rachel: i think Gwen and Svetlana might do it on friday if they are not too tired

me: also, am i driving you or are you going w/ Gwen?

Rachel: i'm going with you if it's an option, although if we don't have enough cars i might get stuck driving

Rachel: my car will be a hot mess if it snows though

Rachel: ps if we have to take Svetlana i will kill her

Rachel: lets take Miranda

me: we should have a draft where each of the drivers is captain and picks who rides w/ them

Rachel: haha

Rachel: idk if people are going to want to ski or not

Rachel: i think Miranda said she'd want to do a half day probably

Rachel: and i do too

Rachel: unless the weather is bad

Rachel: or we are dead

Rachel: but i think if we're going all the way up there we should do 2 days

me: if we are dead, all bets are off

me: I may not ski at all if I'm dead, but that's just me

Rachel: don't be a pussy

Rachel: why will you be dead

Rachel: you're on ECT elective

Rachel: you don't do anything

me: i do lots of things you don't even know about

Rachel: i know everything

Rachel: especially after i break into your phone when you're not looking

Rachel: btw if you'd give me your password, it would be the path of least resistance

me: and with that I bid you adieu

Rachel: night

20

The Irresistible Lure of Delivering Chinese Food

Scheduling dates with Jessie was a challenge because of our competing schedules. In fact, our relationship progressed at a snail's pace in part because we couldn't find a free night between us and in part because we were both passive daters. By nature, neither of us was comfortable being the one to push the relationship forward. About one month after meeting her, I realized that I hadn't even really kissed her — not more than hello or good night. *How odd,* I thought. Still, we marched forward steadily and kept planning our every-other-week dinners, movies, drinks at the bar. I wondered if this could be a person I might end up with. As our thirties approached, a lot of friends were already getting engaged, and I could sense that our cohort was approaching the next stage of dating life. Expectations about relationship viability were beginning to morph toward the long term.

Jessie checked all of the boxes of some idealized list of what a future partner would offer. She was pretty and kind. She spoke adoringly of her family. Her father was a doctor, too, so we had a kind of shared culture around that. Though we were both physicians ourselves, coming from a family of physicians meant it was practically in our blood. More important, her smile made me feel safe, and she was wickedly smart and hardworking. I had a sense that she would be a wonderful mother someday because she was naturally responsible and caring. I thought that her patients must adore her for just the same reasons.

"She's pretty great," I said to Rachel, who was eating ice cream on my sofa, eyes glued to my oversize flat-screen TV.

"Hm? I'm watching this. Shhh."

Rachel had started coming over some nights ostensibly to watch a cable show she didn't have access to at her own apartment, but she was beginning to make herself awfully comfortable on my couch.

At the next commercial break, I repeated my comment.

"She's pretty great — Jessie."

"Who's that?"

"The woman I've been seeing. She's an attending."

"In Psychiatry?" she asked, her eyes widening.

"No. Palliative Care."

"Oh, that's not so scandalous, then."

"Kind of a turn-on for me," I said.

"Well, good for you."

She turned sideways and put her feet on my lap.

"What am I supposed to do with these?"

"You could rub them," she replied.

"You're on my sofa in your sweats, eating my ice cream, asking me to rub your feet. Are you my wife now?"

"You wish," she replied.

We both stared ahead at the TV, which showed a zombie horde attacking a city.

"That thing with the guy from Medicine didn't go anywhere," she said matter-of-factly.

I wondered if she was communicating something unspoken to me, but I chastised myself for falling for it again. *She is not interested in you,* I reminded myself. *If something was going to happen, it would have happened in Mexico!*

"I need more vanilla. Hang on."

I stood up and walked to the kitchen just to get myself some space.

"Hey, can you bring me the rainbow sprinkles?" she called.

"Sure thing, wifey."

"Gross."

On what may have been my sixth or seventh date with Jessie, I noticed myself feeling easily frustrated throughout dinner. When she said that after the meal she had to head home because she had an early schedule the next day, I decided to speak up about my concerns.

"Jessie, have you noticed that our relationship is going kind of . . ."

But I didn't have the right words. I wanted to say it was painstakingly slow—the kind of pace that made me want to jump out of my skin—but I didn't want to seem like I was owed anything, and I knew it was as much my fault as hers.

"Kind of what?" she asked.

"You know, just—not slow exactly—just sort of not as . . ." I hit the wall again. "Are your relationships always like this?" I asked.

She was taken aback by the question.

"Hm, I haven't really had a lot of relationships."

"Really?"

"Why? Have you?"

"I've had some. I mean, I had a serious girlfriend before residency. Eliana."

"From college?"

I nodded.

"I even lived with her for a couple of years during medical school. That's where Magoo came from."

"Magoo the Great Listening Pig?"

"Guinea pig, but yeah."

"I skipped a grade, and then later another grade, and then I did this accelerated program that combines college and medical school, so I was never really the same age as everyone around me. By the time I was in residency, I just felt like an outsider. I pretty much didn't date at all, but at least I had my work to dive into."

"Wow, so you've really never been in a relationship."

"Not if you don't count Joey from the sixth grade. Is this not how it usually goes?"

"I'm definitely not an expert," I said. "But I do think there's usually a little more early-romance passion, maybe even some sleepovers."

"Sleepovers? Oh, I did do those in the sixth grade."

I fake laughed.

"Okay, message received. I really can't come over tonight, but next time I—"

My face must have dropped because she paused and did some mental calculations.

"Or I can make tonight work."

"That would be great. Check, please!"

I fake motioned to an imaginary waiter as a joke.

• • •

The next time Rachel was over, I couldn't help but give her the update.

"It was a lot better to know that she was just not really experienced in relationships; that she was open to meeting me halfway was encouraging."

Rachel stared at the TV. This time she put her calves on my lap, and instinctively I began rubbing them.

"I don't know that we have much chemistry, physically. Not yet anyway, but I hope—"

"Hey."

"Hm?"

"I don't really want to hear about it."

I was surprised. This was the same woman who had wanted regular updates on my online dating.

"I'm just watching this. You can tell me some other time."

In the coming weeks, I tried even harder to get some momentum with Jessie, but her schedule made it difficult. One night she said she would come over around eight o'clock and make lasagna. I didn't like lasagna, and in fact I'm a particularly picky eater for an adult and don't like many foods that the average grown-up enjoys. Despite this, I was so eager to spend time with Jessie that I agreed to the plan without offering up an alternative. If the relationship had any chance of succeeding, it needed to achieve lift-off at some point.

Eight o'clock came and went, followed by nine o'clock. Eventually, she texted.

> Sorry, got slammed with the last patients of the day. Running so late! Want to reschedule?

I knew that the answer should have been yes. She was obviously stressed and tired, and I had already grown irritated by my inability to see her with any regularity, exacerbated by the latest ego blow of her being over an hour late.

No, just come.

She finally arrived at 9:45. She was starved, but at that point I had already eaten something despite trying to hold out, so I didn't have much of an appetite. She spent almost another hour making the lasagna while I continued to stew in the next room. I would have made dinner for her, but she had expressed her desire to be the one in the kitchen making a delicious meal. She had told me that her abilities as a chef were something she was proud of.

Again, I felt eerily like a member of an old married couple. With Rachel, I had been assigned the role of doting husband, while now Jessie was slaving away in the kitchen for me for a meal that I wouldn't even appreciate. The whole situation was made even worse when the lasagna was done. It seemed fine to me, but I just wasn't that hungry and there was *a lot* of it—enough to feed a group of twelve, I estimated. Jessie kept asking if I didn't like it.

"Honest to goodness, it's fine. I'm just not that hungry and it's getting late," I said.

"You want me to go?"

"You could stay," I replied.

"I should go."

My face must have fallen again because she jumped in to try to save the moment.

"But I'm going to make it up to you."

"There's nothing to make right. The lasagna was great!"

"No. I'll tell you what. Let's get together on Saturday. I have no work scheduled. I'll come at a reasonable hour and make my special chicken. It's always good. I promise."

"Okay, it's a deal. I know this is my own neurotic thing, but is it white-meat chicken? I really don't eat the dark stuff."

She looked at me like I was certifiably weird, and surely I was, but why not tell her in advance about a neurotic preference?

"Yeah, I can make white meat," she said, rolling her eyes.

"Great. It's a date."

She left, and almost immediately I got a text.

I'm bored.

It was Rachel. She was on call. I logged on to chat.

me: Aren't you covering the emergency department? How can you be bored?

Rachel: Colleges are on break I think. No heartbroken co-eds reporting with suicidality. I'm all caught up and there's nothing to do

me: Go to bed

Rachel: I want Chinese food

me: I hear they deliver that

Rachel: I didn't bring my credit card

me: Why would you go to be on call without your card?

Rachel: Why would I need my credit card on call?

me: For things like Chinese food

Rachel: Just bring it to me

me: What?

Rachel: Please

me: It's like 11PM. You want me to bring you chinese food?

Rachel: Yes

I knew that it was beyond the realm of normal friendship to bring Rachel a midnight snack when she was on call, but I was drawn in like a moth to a flame. I couldn't seem to resist. I put in the order and began what I was sure would be a quick trip to the hospital.

As I approached with my Chinese food bag in hand, I saw two familiar figures outside the hospital. First was Slippery Nick, who gave me a knowing smile.

"Hey there, Nick."

"Doc. Got anything in there for me?"

"I think they threw in an extra eggroll."

"Yuck. No, thank you," he said, pretending to gag.

"Suit yourself," I replied, continuing along the sidewalk.

The second person I came across just outside of the hospital entrance was Rebecca.

"What are you doing here so late?" I asked.

"My aunt is in the hospital. The nurse just kicked me out so she can rest."

"Oh, hope it's nothing too serious."

"She'll be okay. Why on earth are you here? You're not on call."

"Long story," I replied.

Though it really wasn't. *If any opportunity presents itself to be with Rachel, I seem incapable of passing it by.*

"Are you a Chinese-food delivery boy? I've heard of moonlighting but never like this."

"Yeah, you know how tough it is getting by on a junior resident's salary."

We stood silently sizing each other up on the sidewalk, dimly lit by streetlamps.

"Well, I'll see you," I said. "Hope your aunt feels better soon."

"For the record, I know you're up to something weird here, but I'm too tired from being on the consult service all week, so I'm just going to let it go."

I smiled and passed her on my way into the building. I met up with Rachel in the call room, hoping I wouldn't run into anyone else I knew.

"That took too long. I'm not giving you a tip," she said.

I handed the bag to her.

"Still quiet?"

"Don't ever say that word while I'm on call."

Residents have always been extremely superstitious about jinxing their good luck during a quiet call shift.

We ate and gossiped. I told her about the latest snafu with Jessie.

"If we ever get married, I'm not going to eat white-meat chicken for the rest of my life," she noted matter-of-factly.

We shifted gears and gabbed about residency issues—whether or not the new 4 South attending, Dr. Fennington, was falling asleep during rounds when he would put his hand over his eyes "to think deeply," as he put it. We debated whether Miranda's new boyfriend telling her he didn't "believe in psychiatry" was an immediate deal breaker. She asked if I thought Erin had seemed off lately, and I answered broadly without disclosing what I knew about her recent pregnancy. At the end of our meal, it was time for the fortune cookies. Mine said, *One's actions speak with louder echoes than one's words.* Rachel's read, *A foolish man listens only to his mind or only to his heart.*

"This must be for you," she said.

"Why?"

"I'm not a man," she replied.

I shrugged and reached for the last cookie.

"That's mine," she said.

But I had grabbed it and swung my hand out of her reach. I held it up as one would taunt a small child. She lunged toward me and chased me in the direction of her call room bed. My heart raced as she wrestled me for the cookie, eventually converting her grasps to tickles. By the time we stopped cackling with laughter, she was almost on top of me.

Kiss her, you idiot.

Another moment passed, and she began to pull away.

Do it.

Buzz.

It was her pager.

"Nineteen-year-old home from college. Suicidal."

"I guess it's time for me to go," I said.

"Thanks for the Chinese food."

"This is yours, I think."

I handed her the third fortune cookie.

She smiled as I walked out the door.

Rachel: i'm so excited for next weekend. Time to celebrate STILL BEING IN MY 20s for one more year!

me: have you decided what the plan is

Rachel: no

Rachel: i think if it's nice out we should go to dinner and then just drink on my roof

Rachel: but idk what to do if it's gross out

me: i like that idea

me: not too too much to do if it's gross except bar/club

Rachel: yeah prob bar

Rachel: but like, if it's nice and we're going to come here, then we should eat around here

Rachel: if it's gross, we'd probably go to a bar in back bay or something so it would make more sense to eat out there

Rachel: so i can't pick the place until i see the weather

me: that's a good plan, though. how many people are you inviting

Rachel: idk

me: should be fun however it turns out

Rachel: Yeah, cannot wait. So need this.

21

Just Needing Some Air

A residency class that sticks together for four years forms a special bond. The connection begins when the group starts out, wide-eyed, eager, and naïve. It takes shape early in the trenches of caring for patients together. There is a camaraderie forged through time as classmates become friends and watch one another grow in their capabilities and personal development. As we entered, we felt like children playing dress-up, and by the time we would leave, we had been told we would be psychiatrists. We learned about one another in the ways in which we cared for one another and for one another's patients. The treatment of a patient on 4 South, for example, was a round-the-clock effort. If a resident took care of the patient all day but the overnight call person made a bad decision, the consequences would be awaiting the first resident in just a few hours. If a daytime resident didn't do a solid job admitting a resident, the overnight person might be royally fucked.

"Look, I know you are new here and you are still learning," Gwen said in Feelings class. "But that was completely unacceptable work."

Her ire was directed at Svetlana, who was not one to take a punch without hitting back.

"It is not my fault that you wrote inappropriate prescriptions for a patient," she retorted angrily.

"Let's take a step back because I'm not sure we're all clear on what the situation is," Jen, the new course leader, said.

"It's pretty simple," Gwen began. "Svetlana admitted a patient to 4 South and included just about no useful information. When I came on call, the sign-out was not done yet—"

"I was still working on it. We got slammed with three admissions in the last hour!"

"And so I ended up writing for a number of controlled substances, completely unaware that the patient had been admitted for abusing those same drugs."

"I don't see how this is my fault," Svetlana replied.

"There was no sign-out! That's how we know what is appropriate overnight, and it wasn't there. You set me up to look like an idiot! When Fennington came in, he essentially called me a moron."

"Okay, let's take a pause and try to dissect the complex systems in play that led to this unfortunate outcome," Jen said.

Together, Jen and Nina led us down the path to identify all the ways that our delicate, interdependent coverage system could lead to errors. There was inadequate time to do good work in the late afternoon. There was the pressure to finish up. There was Svetlana being new and thrown into the deep end. There was no way for Gwen to adequately check the patient's history for red flags regarding the drugs she was prescribing. And finally, there was the chronic sleep deprivation that we all suffered, which made us perpetually irritable and unempathic when the job required us to be patient and compassionate.

"It's almost miraculous you guys aren't always at each other's throats, and that's a tribute to what a special group you are," Jen concluded.

Though I had been skeptical about her late arrival into our safe space, Jen had proved herself time and time again in the short while we had known her. The temperature of the room had dropped dramatically in just a few minutes, and we were better able to empathize with one another and brainstorm ways to improve the system. By the end of the class, we were in such a different place that Svetlana began revealing parts of her story we had not heard.

Our mysterious Russian had transferred into our program, at least in part, as an escape. She was fleeing an abusive ex-husband who was still managing to disrupt her life even from seven states away.

"He calls nonstop. He texts me threats. I don't know what he's going to do. He's disturbed enough that I worry. Sometimes I feel like I just cannot . . ." She stopped to gather her composure. "Harvard Longwood is a great program. I was lucky that a spot opened up when it did, but for me it's also a new life, and this guy from my old life will just not let me escape."

Later that week I was sitting with Svetlana on 4 South working on progress notes as the texts came in. They were brutal, mean, verbally and emotionally abusive. The level of vitriol was shocking.

Dr. Fennington popped his head into the room.

"Is there a sushi place around here? I try to take each of our residents out to lunch at least once while they're here."

It was a nice gesture, and as the three of us headed toward the chosen restaurant, I thought it might be a good distraction for Svetlana. I was wrong. The texts kept coming. She silenced her phone, but she knew what would be waiting for her when she reengaged. We ate in mostly awkward silence as I tried to engage Dr. Fennington about his career before coming to Longwood. I noticed out of the corner of my eye that Svetlana was breathing heavily — abnormally heavily.

"You okay?" I asked.

She nodded.

"I just need some air."

She stepped down from the booth and quickly exited the restaurant.

Dr. Fennington paid the bill as I caught up with her. She was sitting on the curb trying to catch her breath. Her face was flushed, and she seemed to be panicking.

"I think it was one of those weird fish toxins," she said.

It was possible that she was having a reaction to the sushi she had just eaten, but I couldn't help but wonder if it was a panic attack. I looked to Dr. Fennington to talk her down, but he seemed completely uninterested in solving the problem. We got back into his car, and I watched from the backseat as he completely disengaged from the obvious crisis happening just to his right.

"So, uh, what are we going to do about Svetlana?" I asked dumbly.

She continued to pant. Dr. Fennington shrugged.

"I guess if you're not better by the time we get back, I'll drop you at the emergency room," he said to her.

And that's exactly what he did. I couldn't believe he had abandoned us.

"Well, back to the unit," he said casually.

"I'm going to stay with her for a bit."

"Okay. Good luck. Feel better, Svetlana!"

I stayed by Svetlana's side until the ER resident, whom we knew well from being on call, came by and examined her.

"This looks like a panic attack," he said dismissively.

"It's not a panic attack," she growled. "I am a psychiatrist. I know what a panic attack is!"

"Well, let's draw some labs and see what we see."

He walked away, and I told Svetlana I'd be right back.

"She's one of us, ya know? Try to be — I don't know — nicer about it."

He looked at me like I had asked him to give her a foot massage.

"Okay, man, but we're real busy down here. If the labs come back normal — and they will — and if she's still oxygenating well, then she's out of here."

I hung around with her for another hour, and her symptoms did steadily improve. All the while I was getting pages that admissions were coming in, even one from Fennington himself, completely oblivious to where I was or what I was dealing with.

"I've got to head back up there," I said to her. "Are you going to be okay?"

She nodded.

"Okay, just text me if you need me."

When I got back up to the 4 South resident room, I was greeted by Rachel with a smile.

She was literally doing a song and dance about it being her birthday.

"How's the day going?" I asked.

"I can't wait for cake tonight! Are you coming?"

"Of course. Wouldn't miss it."

And then I realized that I'd better check the call schedule.

"Oh no," I said.

"What?" Rachel asked.

"You're on backup call."

"So? Svetlana's fine. I saw her this morning."

"She's down in the ER."

"What! Why?"

"Um. She can tell you the story later, but that means that unless she gets discharged, like immediately—"

"My party's off," she said despondently.

We looked at the clock. The call shift was starting in a little over an hour. Even if she were discharged in time, it didn't feel right having Svetlana take the overnight call after the day she was having.

Rachel looked deflated.

I texted Svetlana.

"I'm taking your call tonight," I said to Rachel, and showed her the phone.

Her face lightened.

"You're going to miss my party," she said sadly.

"Better me than you, right?"

"Thank you. I owe you one."

She hugged me.

• • •

Several hours later, while I was slaving away in the ER bunker, Svetlana walked into the room.

"I'm free," she said, looking a lot better than when I'd left her.

"How are you feeling?"

"I'll survive. Thank you for being there with me."

"I'd like to say it's my pleasure, but you are definitely paying me back for this call."

"I could take over now—"

I shook my head.

"Go home. Get some rest. You'll return the favor some other time."

She began to walk toward the door.

"I hope your ex lets you go to live your life," I said.

"Me too," she said.

Just as she reached the door, she paused.

"My thyroid was off."

"What?"

"When they drew labs. They said I had a thyroid storm. That's why my heart was racing and I couldn't catch my breath."

I had no idea if she was telling the truth, exaggerating it, or simply saving face. Even psychiatrists experience stigma around mental health, maybe more than most.

"Well, I hope you get some relief there, too."

"Thanks, Adam," she said as the door closed behind her.

For just a moment, I let myself imagine Rachel eating her cake, and looked up at the patient board. There were four people waiting to be seen and six awaiting placement.

22

Some Kind of Chicken Sociopath

I spent time checking in with each of my 4 South patients the next morning as was customary before heading home after an overnight call shift. I was just about finished when I spotted someone familiar sitting in the dayroom staring at a plate of untouched food. I watched as she lifted a forkful toward her mouth and forced it in, swallowing with a grimace.

It looked like Jane but different. This woman looked healthy. There was no gauntness in her face, no blond fuzz along her cheeks. Her clothes fit without hanging off of her. Did Jane have a sister? I walked slowly toward her, and her eyes met mine. She swallowed quickly as if she was ashamed to be eating. I smiled as I realized it really was her. It was so good to see her looking so well.

"I thought you'd have quit by now," she said.

"Can't get rid of me that easily."

"Me neither, apparently."

I wanted to tell her she looked terrific, but I knew that was a trap. Any comment about appearance can become warped to reinforce whatever emotional and physical distortions the patient with anorexia may have.

"What are you doing here?" I asked.

She paused and took a bite of food as though even that were less appalling than answering my question. Finally, she revealed the underside of her left forearm. It was bandaged vertically about eight inches in length.

All I could think to do was nod slightly in acknowledgment. I hoped she didn't notice my jaw clench and my eyes narrow. Or maybe I hoped she did. On the inside, I was fuming. She had finally achieved a normal weight and, judging from the length and orientation of her gash, had seemingly tried to end her own life. After so many people had tried to help her get to that point, it felt like an insult.

On some level, I knew it was entirely egocentric to think that way. That slash up the arm wasn't at all about me. Maybe it was a result of my powerlessness, but it wasn't done to spite me or anyone else. It was because Jane was still suffering.

"I'm post-call, so I'm about to head home."

"Later, gator," she said with faux nonchalance.

I started to walk out but yearned for a better ending. I paused mid-stride, scanning my mind for another way to engage. Surely, there was something therapeutic I could dig up — but nothing came. My brain was fried from being up all night, and most of the muscles in my body ached. I glanced back at her over my shoulder. She was still staring at me.

"You look like shit, Doc. Go home already."

I nodded and walked out.

When I got back to my apartment, I passed out on the couch almost immediately. I slept for several hours and woke up feeling even worse. My eyes felt like they were on fire, and my back hurt from sleeping on lumpy cushions. I picked up my phone, and there was a text from Jessie saying she would be over in thirty minutes. I had completely forgotten it was the day of our make-up chicken date.

I tried to make myself presentable and get myself in the right mind-set for company, but I couldn't shake the irritability I always had post-call. It was made worse by my interaction with Jane, whom I couldn't stop thinking about. I so wanted her to be well for once, and there was nothing I could do to make it happen.

When Jessie arrived, I put on a good enough show for an hour or so. I'm not sure that she knew anything was wrong until she served the chicken with a big smile on her face and saw me glaring back at her.

"What is it? What's wrong?" she asked, looking almost terrified.

I paused uncomfortably, now glaring straight at the chicken.

There are inevitably moments in relationships when neuroticisms announce themselves. If a relationship is strong, they're perfect opportunities for growth and better integration of two individual people. Eventually, long-term couples develop their own shared idiosyncrasies, maybe even taking a certain pride in harboring them together. This relationship was *not* on strong footing, and I didn't have much emotional reserve at that moment.

"It's just . . . well, I just —"

"What?"

"Do you remember what I said about only liking white meat?"

"What?"

"Remember? I said that I really only liked white-meat chicken, and this is dark meat."

I sounded deranged, like some kind of chicken sociopath, and that's how she looked back at me.

Her eyes began to well with tears, and her lip began to quiver. A good quasi boyfriend would have been sympathetic, but I couldn't muster it, even halfheartedly.

I offered an unenthusiastic hug and felt her tears start to moisten my shoulder. At that moment I could imagine the script of what a caring guy would say. I should have explored her feelings and apologized for my own neurotic shortcomings. We should have talked about what this issue had meant to each of us and how to overcome it or at least found sympathy for each other's positions. Why couldn't I do it? She even started the process for me.

"I'm so embarrassed. I don't mean to be crying about this."

She looked up at me with big, sad eyes. The warmth that she exuded constantly had morphed into a kind of naïveté that I had no patience for in that moment.

"It's just that I've always thought food is kind of a core part of two people being together in life, and now I've tried to have that with you twice and failed twice."

She lobbed a slow pitch right over the plate, and on a better day I would have knocked it out of the park. I didn't even foul it off. I just let it go by.

"Aren't you going to say something?" she asked.

I should have said something kind, but I didn't. I could have even replied with a banal excuse about just being sleep deprived that would have at least saved face for us both, but I didn't.

"This isn't working," I said.

She recoiled and looked at me, completely betrayed. There

was *still* time to salvage if not the relationship then at least our dignity, but I let that pass, too. I simply didn't have it in me.

"I think I should go," she said.

"Okay."

After packing up her chicken and everything else that she had brought, she left the apartment and the door slammed shut. I curled back onto the couch and felt nothing.

When I returned to 4 South two days later, I scanned the unit for Jane. There was no sign of her. I looked up at the patient list at the nurses' station. Her name wasn't there.

Gangly old Fennington wandered by the doorway with his head buried in some papers.

"Dr. Fennington," I shouted after him.

He didn't hear me, or maybe just ignored me and kept on walking.

"Dr. Fennington!" I shouted as I jogged up the unit's main corridor after him.

Several patients stopped what they were doing to look up at the commotion I was causing. Finally, I caught up to Fennington and placed a hand on his shoulder.

"Oh, hi. Alex. Hello."

"It's Adam."

"Adam. Of course."

"Do you know what happened to Jane?"

"Who?"

"Young woman with anorexia nervosa. She was just admitted at the end of last week after a suicide attempt."

"Oh yes, I discharged her."

I was stunned.

"Her suicidality seemed to have resolved, and you know we're not a hostel."

"What?"

"We can't keep people on the unit any longer than they need to be. As soon as we think they are out of immediate harm's way, it's time for them to move on to their next phase of care."

I was deflated, and I didn't even know why. Did I really think more time on 4 South was going to help Jane get better? The more time I spent on 4 South, the less well I seemed to feel.

"You can't save 'em all, Alex."

With that final insult, he returned his attention to the papers in his hands and began wandering down the hall once more.

Rachel: what do you want to do

me: i would like to drink and am otherwise fairly flexible

me: food/movie/bar/be outside — any of those would work for me

Rachel: i am starting to get hungry

me: what about that sushi place — Kyufuga?

Rachel: eh

Rachel: it's expensive

Rachel: and not really THAT good

me: are you in the mood for anything in particular?

Rachel: idk

Rachel: will you eat pho

me: what is that

Rachel: it's kind of like a soup

Rachel: with noodles and chicken or beef

me: yeah, i eat that

me: where do you get it?

Rachel: at a place that sells pho

Rachel: there is one near my house

me: okay

me: do you want to tell me where it is and meet me there or should I come to your house

Rachel: just come here

Rachel: idk if there is parking over there

Rachel: meters are free after 8

me: ok, come nowish?

Rachel: yeah

23

Like Free Therapy

After I officially ended things with Jessie, Rachel and I started hanging out almost all the time. We were toeing the line between close friends and something more. By the middle of our second year in residency, we were both emotionally burnt out and physically exhausted. The idea of dating someone new seemed impossibly daunting after my recent false starts, and Rachel and I had developed an easy rapport. More often than not, we got together after work and hung out like best friends *plus*. After the first few movies we had watched together, I began to grow dissatisfied with rubbing her feet and calves while staring at the screen. During one late-night showing, after we had both had a glass of wine, I leaned over a bit and asked if it was okay if I lay down next to her. She assented.

"I promise not to fall in love with you," I said as much to myself as to her.

I had been infatuated with Rachel since the moment I met her, but even after months of more-than-friends behavior, I couldn't seem to get past the potential shame I would feel if I expressed my feelings in earnest only to be rejected. She had even told me that Gwen and Miranda had asked her point-blank if anything was going on between us and she'd denied it, noting that we just liked to hang out.

One night I was out with Svetlana to help her vent about the troublesome ex-husband who continued to make her life miserable from afar.

"Why don't you just ignore him entirely?" I asked.

"I can't," she replied sadly.

"Because?"

She traced the pattern on the napkin in front of her, avoiding eye contact.

"I have a daughter."

I had always assumed that psychiatric training would help me know what to say in complicated, uncomfortable situations, but I was dumbfounded.

"She's with him right now. I need to figure *something* out, for her sake."

Now I was the one avoiding eye contact, wishing I could think of the right response.

"It's okay if you don't know what to say. I don't either, and talking about it just makes me sad."

"Then let's not," I replied.

We were two doctors at Harvard but may as well have been grade schoolers encountering a problem way too big for us to handle. Svetlana put on an appearance of bravado and what-you-see-is-what-you-get, but the more I got to know her, the more I realized how vulnerable she was. I had actually had the same feelings about many of the tough overachievers in our so-called Golden Class. In our late-night get-togethers, even Rachel had begun to reveal imperfection, even self-doubt.

I signaled the bartender for another round.

“What about you?” Svetlana asked from atop her five-inch stilettos.

“Hmm?”

“Tell me something better about your life. Any new romance?”

I shook my head.

“I see the way you are with Rachel. Everyone does. Are you guys dating or what?”

I shook my head again.

“We just hang out a lot.”

“But you like her.”

“Doesn’t matter,” I replied.

“Why not?”

“She doesn’t feel the same way.”

“How do you know that? I see the way she acts around you. You are wrong. Remember the holiday party? She had her hands all over you.”

“We were all kind of tipsy at that party.”

“I have heard it said that people only express their real wishes when they’ve had a drink.”

“Maybe Rachel and I would be a couple if we drank more.”

We each paused to sip our beverage.

“That holiday party was wild though. Did you see Erin get cornered at the bar?” she asked.

“No, by who?”

“One of the old bald guys. Full professor type. I don’t know their names yet.”

“She’s so friendly that I think sometimes men — older men in particular — get the wrong idea.”

“When we were all leaving, she was still trapped with him!

I think sometimes she has some responsibility in it, too. She didn't look like a victim when I walked by."

At our next Feelings class, we got Erin's side of the story.

"I was just at the bar, and Randell just sidles up right behind me."

"Randell," I said with a snort. "That guy is such a narcissist."

There was a knowing smile from Nina and Jen.

"What makes you say that, Adam?"

"I was chatting with him after his class last week, and we walk out of the room together and down the hall. Then he takes a left turn, and I realized he was heading into the bathroom, mid-sentence. So I asked myself—am I supposed to follow this guy into the bathroom? I could hear he was still talking, so I did. And then he starts urinating, and he keeps talking. What am I supposed to do with that? So I just washed my hands and looked down into the sink. Then he finished and kept talking as though nothing was weird about that at all."

"Sounds exactly like Dr. Nathan Randell," Nina said.

"But sorry, I interrupted Erin. Go ahead," I said.

As she spoke, she began to turn red.

"Well, you know, it was the holiday party. The alcohol was free."

"Only two drinks were free," Miranda said.

"Yeah, okay, but Gwen and Dana had given me theirs. And Ben."

The room fell into uncomfortable silence. All eyes were on Erin.

"I wasn't in a good place. Bobby and I had just had an awful fight. He was refusing to come to the party—as always—and

I gave him an ultimatum, and then when tempers were really flared, I told him that I had been pregnant and had miscarried."

This was news to everyone in the room. A room full of trainees in the human condition had no idea what to say. Nina and Jen wisely gave her the space to decide how she wanted to go with such a public revelation.

"I told him, and he cried — which he never does — and then we sort of just left it because I was already late to the party. So I went and he stayed behind. Then I found myself at the bar and Randell came up next to me. I told him I had really been enjoying his classes and I looked forward to doing an elective with him next year. He could tell I was upset and just started listening to me. It seemed innocent enough. It went on like that for a while, and at some point I realized I was drunk, and I think *he* was drunk, too! He seemed warm and kind and empathic. It almost felt like I was getting free therapy."

Erin's face transformed. She placed her hand over her brow, averting eye contact.

"Then I started to realize that we were two of the only people left at the party. That was the first time I understood that there was something off about the situation. I suggested that it was probably time for me to go home. He agreed, which made me feel better again, but then he suggested that he give me a lift. At this point, there was pretty much no one left except for the bartender. I had no buffer and couldn't think of what to say."

"You were feeling unsafe? Like he might do something inappropriate?" Jen asked.

"It seems so ridiculous. I told myself, Randell's not a sexual predator. He's like seventy! He could be my father."

"And yet more often than not, it's people in positions of trust that are able to cross those lines without anyone raising red flags about it," Nina said.

"We walked out and took the elevator down to the street. He had hit P1 for the parking garage, and at the last second I hit G and the elevator doors swung open. *I think I'll just grab a cab,* I shouted as I left the elevator without looking back. *Thanks for listening!*"

The room was hushed, waiting for someone to know what to say next.

"I haven't seen him or spoken to him since. It's just so *strange.* He didn't do anything wrong *exactly,* but why did I feel so unsafe?"

"You found yourself in an uncomfortable position that had the potential to become dangerous, but you weren't sure. It sounds like you didn't want to overreact, but you also didn't want to underreact," Jen said.

"And I think what may have added to your unease is the fact that you played a role in how it unfolded," Nina added.

"Exactly! I could have gotten up from that bar at any time. It would have been so easy to just excuse myself. Instead — once again — I let the idea that a man would listen to me and support me and be kind to me lure me into something more than it should have been. I feel guilty. I was complicit."

It reminded me of some of my patients in the way that her actions betrayed what needs and desires were beyond her reach. Certainly Jane wanted to be well, but her own behaviors made it impossible. My new therapy patient, Jim, knew that he was pushing his marriage to a bad place because of his own narcissistic drives, but he couldn't stop because there was some kind of hidden value to acting the way he did. It was protective to his sense of self, something even more powerful to his psyche than maintaining a healthy marriage. Even Charlie, the suicidal patient newly diagnosed with cancer, was so desperate for a sense of control in his shattered world that he would consider deliberately enacting exactly what he feared so he would achieve a sense of agency. To me, Erin's situation was on a different plane mainly because it occurred within the world of our residency training environment. Even the loosened atmosphere of a departmental holiday party warrants proper behavior between supervisors and trainees.

"Are you going to say anything? Maybe to Dr. Redding?" I asked.

She shook her head.

"I don't want to be known as *that* woman."

"What woman?" Jen asked.

"The one who winds up in sketchy situations with male attendings. When they think of me, I want them to immediately recall my skill set, my accolades, my dedication. What would reporting Dr. Randell even do? What could I even say? *He listened to me longer than was appropriate? He offered me a ride home?* No. I need to get my shit together. I need to stop letting these men draw me in. I need to ace this fucking residency so I can be chair of the department someday, and somewhere in there I'm going to have a damn baby, too."

Her fists were clenched. If Feelings class were an actual therapy group, we would have spent some time unpacking the rest of what she had said. As it was, our time was running short and Jen and Nina had to do their best to weave the dilemma back to educational pearls. They had told us in the beginning of residency that even with an unstructured curriculum, Feelings class would always highlight lessons the group could learn from our discussions.

I don't recall what those were, though I certainly do remember understanding—maybe for the first time in my life—the kind of tightrope professional women walk on a daily basis that as a young man I had been shielded from entirely. I didn't envy Erin for having to navigate an apparent parade of senior male supervisors whom she could never let herself fully trust, all while having the drive to conquer the academic world and sit atop of a department at Harvard. Above all of that rested an insatiable need for validation. My pressures felt less intense. For me, I was relieved just to get through a call without messing up terribly, and if there was a chance that Rachel might kiss me, well, that would be enough for me.

24

The Joy and Sorrow of Musical Staircases

"His wife told him she wanted to leave him," I said.

I sat in Meg Mook's office, looking out the big windows behind her desk down to a lacrosse match happening at the high school that neighbored the hospital. Dr. Mook was not formally my supervisor on the case, but I had been getting mixed messages from the faculty members to whom I'd been assigned. One thought I should be more active in guiding Jim's behavior, assigning him lifestyle changes almost as if they were homework, while another thought I should be focused primarily on early childhood. Neither approach felt quite right to me when the patient was coming to me in distress about his marriage falling apart. When I saw Dr. Mook's office door open, I took my chances with some curbside supervision. As the woman in charge of the outpatient clinic, she took the opportunity to engage with residents very seriously and never turned away a helpless trainee like me.

"How was I supposed to respond? I don't even know the first thing about being married," I said.

"Often our patients are living lives that we can't necessarily envision, but there's usually an opportunity to connect with their experience. Have you ever felt fearful that you might lose something important to you?"

"Almost every day of my life," I replied.

"Once you accept the fact that you are probably not empowered to save his marriage all by yourself, you may be able to settle into the work of helping guide him through his experience of a challenging situation."

"How do I do that?"

"Start by reflecting with him about what it must be like to be recently married and on the cusp of divorce. See if he has the capacity to explore those feelings, and be there to hold them with him in the room. That alone will go a long way toward helping him go forward."

"I can do that, but what happens then?"

"Then you see what he brings to the next session and the next and the next, and over time you may even discover some things together about his inner world that can help him become more like the person he wants to be."

"Okay, thanks, Meg. I've got to run. I'm supposed to start my rotation at the children's unit tomorrow, and it turns out I don't know anything about kids. Seems like there are a lot of things I know very little about."

"My door is always open."

Arriving at the children's unit at the neighboring hospital was like entering an entirely different world. Whereas 4 South had a well-earned grit and grime, everything at the children's hospital sparkled. In the lobby there was an immaculate saltwater fish tank. Kids had lined up pointing out Nemo and Dory lookalikes. Walking up the stairs was literally a musical experience, as the main lobby featured a musical staircase that rose in tone as you climbed. By the time I reached the second floor, I was always in a better mood than when I had begun the journey five seconds earlier.

The psychiatric unit was pristine and looked more like some kind of discovery zone play area for children than a mental health ward. There were bright paintings on the walls and new-looking carpets on the floors. There was a game room with the latest video game systems and classrooms on either end of the hall. There was a mini-gymnasium and a room for art therapy. What an amazing physical space to receive care, I thought.

I was met by eager staff members including what seemed like an army of therapists from every methodological school. There were cognitive behavioral specialists and music therapists and art therapists and pet therapists who came and went all day.

The kids on the unit, who ranged from age four to seventeen, seemed much healthier overall than some of the severely mentally ill patients I had seen on 4 South. To me, they just looked like any other kids. I reviewed their medication records and was shocked to see the number of antidepressants, anxiolytics, antipsychotics, and mood stabilizers these children were being prescribed.

The attending who had been assigned to supervise me took me out to lunch across the street from the hospital.

"Notice anything?" Dr. Quadratto asked.

It seemed like a pretty unremarkable restaurant experience to me.

I shook my head.

He looked at me coyly. I had heard *a lot* about Dr. Quadratto —frankly, too much, in my opinion. He was just a few years out of residency and child fellowship training, and all the women in my class seemed to think he was hot. I was ready to hate him for this alone, but within ten minutes of meeting him I noticed myself feeling completely at ease with him. He was good-looking, and he knew it, but he also had a kind of superpower in his ability to make you feel like you were on the same level as him, no matter where you were coming from. I noticed him chewing the fat with teenagers on the unit just as smoothly as he gave out high fives to the little kids. Somehow, everyone seemed to connect with him, and he seemed even more authentic with the kids than with their parents, which I supposed made sense for a person who specialized in working with kids. When he was talking with his patients, he seemed so genuine, unshackled by the societal demands and expectations of what a doctor is *supposed* to be like.

"Take another look," he said. "Anyone seem familiar?"

I looked around again. Almost every table around us featured patients from the unit with their parents eating lunch.

"How are they here? I mean, isn't it a locked unit?"

The idea of patients on 4 South going out to lunch would literally set off alarm bells.

"On the child unit, we give patients passes to go out with their families as they start to get ready for discharge. It's a way for us to ease them back into the world. Kids start with a lunch pass, and if it goes well, they often will go home for a day or even a weekend before being discharged."

In the years since, whenever I see children eating at a restaurant near a hospital, I always think of that policy. It's a necessary reminder that the world is brimming with people suffering through invisible illness.

It wasn't until my second and third day on the unit that I truly began to appreciate the sickness that tortured most of the kids admitted. Rachel had actually just rotated off the service and was over at my apartment one night early in my rotation. She gave me a true accounting of the children's unit.

"Watch out for Trevor."

"The kid wearing the Spider-Man pajamas all the time?"

"Yeah, he's a biter. And never, never, never let Alisha near the door. She will bolt so fast, and she'll be in New Hampshire by the time anyone realizes."

"Billy seems like a sweet kid."

Her face dropped.

"What's up with Billy? I already know about the drooling."

Her sadness turned to anger.

"The drooling is because of the clozapine, which has also made him gain thirty pounds in the last month, by the way."

"Why clozapine? That seems pretty aggressive for a twelve-year-old."

"We tried everything with Billy. Risperidone, olanzapine, every therapy out there. Clozapine is the only thing that quieted the voices. But it's hard to watch him like this. The drugs help, but at a real cost for a kid like Billy."

"It seems too hard."

"It's hard, but it's the only thing I want to do."

"You're going into child psych?" I asked.

She nodded.

"Sick kids, entitled parents, the whole thing?"

"It's all I can picture doing," she said. "If we make a difference when they're kids, we can help them for their whole life."

I admired the passion she had for part of our job, or more accurately a career that awaited her after graduation. Fellowships were optional additional training programs that followed residency for doctors who wanted to subspecialize within their fields. It suddenly dawned on me, though, that Rachel going into a child psychiatry fellowship would mean the end of her being with me at Harvard Longwood, and that time was coming soon; residents are allowed to apply during their third year and fast-track into the next two years of training, which meant skipping the fourth year of our current residency. It was exactly what Rachel intended to do. We were already approaching the end of our second year, and Rachel had a whole list of programs she was looking at, including several in California.

Whenever we would talk about our future whereabouts, she would reference someplace warm.

"I'm so tired of these winters," she said.

The thought of losing her, even as a friend, sent a chill through me.

"What about your family? They're all around here."

"Yeah. That's a real factor, but I can always come back after two years in the sun."

"Okay, enough fantasizing about sunny beaches. Let's put a movie on."

She chose a classic, *The Silence of the Lambs,* but I couldn't get my mind off of the distinct possibility that my time with her was coming to an end. In fact, I couldn't focus on the movie at all.

Like every other night we had hung out, we fixed our gaze on the screen in front of us, but our bodies edged closer together until we were holding hands and lying down on the couch. Somewhere in the midst of Hannibal Lecter donning that straitjacket with protective face mask and Buffalo Bill sewing himself a suit of human skin, Rachel turned back to me. Our eyes met and we finally kissed.

The next day on the children's unit, I was light on my feet and feeling like nothing could touch me. I had admitted a new boy, fourteen-year-old Daryl, who had recently run away from home. There was nothing too remarkable on our initial interview. He had been living with his twin brother and father, whom he wasn't getting along with. He missed his mother terribly, whom he didn't get to see very much after his parents' divorce. On top of it all, he was getting bullied at school. The social worker on our team had already arranged a family meeting, and when I walked into the meeting room, there were two middle-aged people sitting in chairs awaiting our arrival. Though neither was saying a word, their body language was screaming, suggesting that they couldn't stand to be in the same room. It wasn't until I walked in with Quadratto, the social worker, and Daryl that I saw the woman's face. It was Deborah.

25

EXP

"Oh, it's you!" Deborah shouted with a smile. "You're a child psychiatrist, too? Quite the renaissance man."

"You know each other?" her ex asked almost jealously.

"I'm just rotating here," I said. "I still do overnights at the adult hospital."

He lifted his face in recognition, as if to convey that now he understood that I had taken care of her. It was a very odd dynamic. Technically, I could not say how I knew Deborah. It would have been a privacy violation, but she was free to say whatever she liked. While in the family meeting, she kept pretty tight-lipped about how she had been doing and focused on her son. Deborah and her ex had been acting out in front of their kids — arguing over alimony, denying visitation, and just generally exuding rage toward each other.

"Daryl's the apple that doesn't fall far from the tree," the ex said, motioning to Deborah. "He's always been the emotional one. Maybe if she had let me —" As if he could feel her eyes staring at him with fury, he stopped mid-sentence. "We both just want him to be better," he concluded.

"Divorces are never easy on kids, but there are some helpful techniques to make sure Daryl and his brother do as well with it as possible," Quadratto said. "We'll try to focus on some of that here, and it may also take a longer-term commitment outside of the hospital involving family therapy."

He was so smooth. His voice was casual and calm, and even Deborah couldn't stop staring at him.

When the meeting ended, I quickly excused myself. I had to run back to the adult hospital, where I would be taking the overnight call. Deborah said she would walk out with me. Under other circumstances I would have found a way to avoid the interaction, but I was so curious to hear how she had been doing. It had been months since I last saw her going under anesthesia for ECT. She looked so much better now, in all kinds of ways. Her face was brighter and more emotive. She made good eye contact. She looked well.

"The ECT saved my life," she said in the elevator.

"I'm so glad you had a good experience," I replied.

"No, you don't understand. It really saved my life. I would be dead or in a ditch right now if it weren't for that treatment. I still get it once a month! I hate it—I really do—but I keep every appointment because I remember what it was like before."

"That's really great, Deborah. Thank you for telling me."

"Thanks for everything you've done and will do for Daryl. I know he'll be in good hands. And that supervisor of yours doesn't seem too bad, either!"

"Yeah, everybody seems to like that guy."

We had arrived at the exit to the street when she put her arms around me and hugged me tight. When we broke apart, I turned and walked away feeling like maybe all of the struggles of the last two years and my feelings of inadequacy were worth it if I had played even a small role in helping someone like Deborah get back on her feet. She wasn't without stress in her life, obviously, but it just seemed like she was in a much better place to handle it.

My pager buzzed, startling me. The vibration had begun to haunt me with phantom pages as I fell asleep each night. My shift had only just begun, and already I was being called to assess a patient on the medical floors. I called the medical resident back, motivated to see if a so-called curbside consult would be enough to get by — often a quick clarification by a specialist is all that a medical resident really needs, as opposed to a time-consuming full consult. I knew there would be several patients already waiting to be seen in the emergency department.

"We'd like to discontinue the olanzapine. It seems to be making her sedated, and frankly I don't think her blood pressure can handle it right now. She's pretty sick," the medical resident said over the phone.

"Who are we talking about?"

"072915," he replied, citing her medical record number.

I pulled it up on the computer. It was Jane, and my heart sank.

"Hold the olanzapine for now, and I'll be there as soon as I can."

When I cleared the board in the bunker, I made my way up to the seventh floor to see Jane in person. It hadn't been so long — a matter of weeks — since she had looked so healthy and well on 4 South, but now she appeared to be an entirely different person. She was unconscious when I walked in and didn't respond to my knock on the door. She had a tube pumping pureed food into her stomach, and her skin hung from her bones. I looked up at her heart monitor and noticed that she was severely bradycardic—her heart was beating too slowly, a common sign of late-stage malnutrition.

"Jane," I said in a hushed tone.

She didn't respond.

I should have gone back to the bunker to write my notes. There would be new consults coming in momentarily. I could almost sense it. It had been *too quiet* a start to my shift.

Instead of leaving, though, I pulled up a chair and sat next to her. Without my making another sound, her eyes opened, and a slight smile crept across her gaunt face.

"You again," she said quietly.

"How are you feeling, Jane?"

"I'm exhausted."

"They're going to stop the olanzapine. That should help you perk up in the next few hours."

"No, I mean, I'm just so tired of it all."

"I know you are."

She motioned to the nasogastric tube plunging down into her throat.

"Court-ordered."

I nodded.

"I'm losing."

I started to speak, but my voice quivered, and that sensation in my throat made my eyes reflexively water.

"It's okay," she said. "It's time."

"Since the day I first met you, you've always told me exactly how it was all going to go," I said. "I think you may be wrong this time. They're taking good care of you here."

My pager buzzed again. It was Beatrice, the Page Torturer, with a question about her over-the-counter melatonin dose.

"You can go. You'll be okay," Jane said, closing her eyes.

"You mean, *you'll* be okay," I said.

She let out a soft sigh and fell back to sleep.

I walked out to the nurses' station, jotted down a quick note, and headed back down to the bunker. There were three new consultations brewing, but I sat at the desk and called Beatrice back. Instead of trying to answer her question as quickly as possible so I could go back to work, I found myself asking her how she was doing. What was she reading these days? Did she have any pets? How did she like to spend her Sunday mornings?

Eventually, Beatrice hung up with me—something that had truly never happened before—and I returned to the work ahead of me. Three patients with suicidal ideation awaited, along with an old married couple who had the same fixed delusion that their local priest was making false police reports against them.

I saw them all and wrote notes straight through until sunrise. Before leaving the building into thc blinding light of early morning, I checked in the computer system's virtual portal to see how Jane's vitals were. Her heart rate and blood pressure were lower even than when I had seen her hours earlier. Then I put one foot in front of the other until eventually I made it home, and I collapsed into my bed.

When I awoke several hours later, the first thing I did was pull up the virtual portal again and type in Jane's number.

West, Jane 072915 EXP

It wasn't the first time I had seen those three letters next to a patient's medical record number, but it was the first time it had ever stung so painfully. The patient had expired; Jane had died ten hours after telling me I would be okay.

I didn't feel *okay.* I didn't feel anything but a consuming sense of emptiness expanding in my chest. I stumbled through the kitchen, down the hall, and into the bathroom, where I turned the shower on full blast until it was scalding me, and wept.

26

An Office with a Window

There was a period before my time that all residents in psychiatric training underwent extensive therapy themselves. It stemmed from the psychoanalytic tradition. It was thought that one could not be a truly impartial emotional guide for patients without being attuned to one's own experiences and implicit biases. In recent years this tradition has softened tremendously. While residents from many other fields seek out therapy in secret, the process is less stigmatized for psychiatric residents. In fact, our residency had set up a freely shared list of local therapists who took our insurance or offered rates on a sliding scale based upon our relatively low resident salaries. For days after Jane died, I felt lost. It was as though I had spent my entire life on a false path. I had thought that being a doctor would at a minimum leave me feeling as though I had helped my patients even when they had bad outcomes. It felt as though with all of my training and all of my years of study, I hadn't been capable of doing *anything* for Jane, and I wondered why I had devoted myself to this path at all.

I discussed it in Feelings class, and Nina and Jen reminded me that there were resources for this kind of professional crisis of confidence. I didn't think I needed therapy, but still I scoured the provider list hoping that somehow a name would jump off the page. Disappointingly, I only saw names of people I knew personally and therefore would never go see as a therapist, or people I had no idea about. I decided to take the list to Dr. Mook, who ran the clinician health service in addition to her many other jobs.

I sat down with her and told her what I had been going through. She listened intently and expressed empathy for what I had experienced — the feeling of powerlessness I had over Jane's course. She paused and thought for a while. She asked if I had a gender preference or if I wanted to see someone old and experienced or younger and more relatable to my current circumstances. Finally, she came up with the name of a woman who had trained in our program ten years earlier — Katherine Pettyjohn.

"I think that would work. I think she would be a good fit," Meg said.

Dr. Pettyjohn's practice was just a few stops down the road on public transit, and she took my insurance. If Meg thought she was good enough for me, or maybe vice versa, then I owed it to her to give it a try. I called her up, and we scheduled our first appointment for the next week.

When I arrived, I sat in the waiting room nervously. I hadn't felt ashamed about asking for help until that moment. There was a bias in my mind that if I were a psychological guide for my patients, needing my own guidance would expose me as a fraud. I felt that at any moment a colleague would step out and spot me. They wouldn't point a finger at me, but quietly they would judge me for being at a therapist's office. I thought about it some more and realized that it would also mean *they* were in a therapist's office. I was able to shake the shame pretty well, even when several weeks later I did see one of the senior residents exit from the office just as I was entering. We simply smiled at each other and went on our way.

Entering from the waiting room to Dr. Pettyjohn's little office was like settling into a warm bath. It was nicely decorated with homey art on the walls and perfectly sized throw pillows on the couch. She had a cup of tea next to her chair and offered one to me as well. The windows were enormous, and I thought how nice it would be someday if I could have sessions of my own in an office like this.

When she asked what had brought me into the office, I winced at the familiarity of the interaction. I told her about Jane.

"I'm so sorry," she said.

She was wearing a cotton dress and comfortable-looking shoes, which I continued to stare at.

"I'd like to hear more about your connection to this patient, but I'm also picking up on something else that I can't quite put my finger on."

"What do you mean?"

"Well, as you know, there won't be anything that I can do to help with Jane, though together we can hopefully find some ways of healthy grieving and coping with loss."

She paused and leaned just slightly forward.

"There's something else, though. Right? What brings you isn't the idea that you need help grieving."

I shook my head and began to tear up.

"I don't know that I want to do this anymore," I said, sniffling.

She pushed a box of tissues in my direction.

"Go on."

"I'm working so hard at it, and when I actually sit down to think about what is important to me, I don't even know if I want to be a psychiatrist."

"What is important to you?"

"I want to — I need to — well, I don't know really."

"Take your time and tell me what comes to your mind. There are no wrong answers," she said.

I blew my nose and threw the tissue into the trash bin to my left.

"I want to help people. It's a cliché, I know, but I think it's true for most people who go to medical school. I liked the idea of being a doctor because it seemed almost like the perfect combination of doing well and doing good. It pays well — not yet, really, but hopefully someday — and I could feel good about making a difference in people's lives."

"And it feels like you're not achieving that?"

I shook my head.

"Not even with Jane?"

"Of course not. She's dead," I said flatly, even though the anger was beginning to rise from within.

"Death is a hard thing to process. In medicine it's often a negative outcome despite its inevitability over time. But I'm not sure that means you didn't make a difference in Jane's life. Are you?"

For a moment I was speechless, pondering the question. Though Jane had shown me a very gruff exterior in our early meetings, she also let me in to share in her pain. In many of our encounters, she had seemed almost thirsty for someone to simply be in it — all of the suffering and the anguish and rage — with her. That was a function I could serve.

Pettyjohn and I went on to talk about a number of positive and negative experiences in my training. Upon closer examination, I realized that there were many patient cases where even though I was learning as I went, I could see signs that my actions were helping distressed people. There were others that made me feel helpless, enraged, unprepared, and even depressed myself.

"It almost sounds like projective identification," she said.

"Oh, yes. Right," I replied, not knowing that term.

"Sometimes when patients experience intense emotions in themselves, they can send them out to the world, and the people who are most receptive — in our case, psychiatrists — can sometimes start to feel the very same intensity of emotions that has been put out without even realizing it."

"Is that what's going on with me?"

She shrugged.

"I don't know," she said casually. "We can try to find out if you like."

Pettyjohn and I met weekly for about two months. It was what would be considered a brief psychotherapeutic intervention. Grounded in insights, Pettyjohn offered a supportive stance and environment for me to come to terms with issues on my own. I found the sessions enormously valuable in helping me to see the field, and my experience in it, with greater clarity, warts and all. As I transitioned into my third year, which would be primarily in the outpatient setting, we decided that it was a good time for me to stop coming in.

"I'll be here if you'd like to get restarted in the future," she said.

"Thank you," I said.

Rachel: are you excited that i decided we're sharing an office next year

me: i was surprised to see my name signed up with you without prior discussion

Rachel: we only overlap on the outpatient side on Thursday mornings for the 3rd quarter of the year so it's a nearly perfect fit

Rachel: obviously i will use the office during that time and you can find another place to be

me: fine but i'm going to refer all my most difficult patients to your psychotherapy group

Rachel: i'm sure with you as their therapist they'll need the extra help

me: how true. Ouch, but how true.

Rachel: i'm going to decorate

Rachel: i hope we get a good office

me: can we paint it pink?

Rachel: i'm thinking pink and yellow

Rachel: stripes

me: i love yellow!

Rachel: how is it going at children's

me: fine, Quadratto is so good.

Rachel: fyi no one calls him that. It's either Dr. Q or just Quad.

me: good to know, also it feels right to introduce myself to the kids as Adam, one of the doctors here. is that appropriate?

Rachel: idk

Rachel: i went with dr. Zeiglerheim

Rachel: but no one could figure it out

Rachel: so they all ended up calling me Rachel

Rachel: esp b/c the staff referred to me as Rachel to the pts

Rachel: i think Ben was dr. Ben

Rachel: you could do dr. adam

27

Friends Don't Have Sleepovers

Finally, Daryl's planned discharge day had arrived. I had been looking forward to it and hoping it would come before I officially finished my rotation at the end of the week. We had a family meeting scheduled for that afternoon. Daryl and I were having our final session playing Jenga in the dayroom.

"How are you feeling about going home?" I asked.

He shrugged.

"It's okay to feel nervous. You've been here for a while, and it's normal to have all kinds of feelings when you go back home."

"I'm excited to see my brother," he replied.

His face became crestfallen.

"What is it?"

"I wish we could just live at Mom's."

"You split time there and at your dad's, right?"

He nodded.

"It must be hard having two homes."

"It's not that," he replied.

"Then what is it?"

"At my mom's I'm comfortable. I can be myself. At my dad's I can't."

"What do you mean?"

"I can't joke around or mouth off or anything, but even when I don't mean to, sometimes he thinks I'm making fun of him or something and then I really get it."

"What does that mean?"

"With the belt."

He looked at me like I was an idiot, and I felt like one.

"Your father hits you with a belt when you make jokes?"

He nodded and made his next Jenga move, but the tower collapsed.

"Sometimes I think they just think I'm depressed because I'm trying so hard just to blend in. Never can for very long though."

I exhaled loudly.

"That sounds really hard."

He shrugged, and we sat quietly for a minute.

"It's almost lunchtime," I said. "I'll see you at the family meeting later."

I immediately paged Quadratto, and we met in his office, which was decorated with all kinds of toys and memorabilia. He said it made for evocative conversations with his patients, but I had a feeling he genuinely enjoyed his surroundings.

"Apparently, Daryl's father hits him with a belt."

"He told you that? Today?"

He let out a sigh and turned his body to the computer, but his head still faced me.

"What did he say, exactly?"

"He said that he's not comfortable at his dad's house because if he lets loose at all, his father hits him with a belt."

He pulled up a website from his browser and clicked into a document.

"You ever fill one of these out?" he asked.

"I don't even know what it is."

"It's a 51A. As doctors, we're mandated reporters. Whenever we hear about a child in danger, we have to formally file with the Department of Children and Families."

"Oh my God. Should I have said anything? What if they ruin this kid's life?"

"First of all, you were mandated to not only say something but to formally file. Here — start filling this out."

He handed me the keyboard.

"Second, if this is the first filing, it's highly likely that they will just screen it out. They are not going to do anything drastic without a mountain of claims and evidence. Frankly, it can be the absolute shittiest part of this job when you hear about more than just a belt and know nothing is going to change."

I began typing away about what Daryl had revealed.

"Document it all, word for word, as best you can, with no literary nonsense, exaggeration, or minimization. Just tell it like it is, and then this afternoon we'll tell his parents that we submitted it."

"Are you kidding? We have to tell them?"

"Of course we have to tell them. Don't sweat it. I'll be there with you."

That afternoon Deborah and her ex-husband were already seated when we arrived. Daryl was sitting in Dr. Quadratto's chair, twirling in circles.

"Making yourself at home, I see," Quadratto began.

"How do you do this without getting dizzy?" the boy asked.

"The secret is to always stop on exactly the tenth turn," he replied, grabbing the chair. "Go sit between your parents."

"This the kind of so-called therapy my son has been getting here for the last God knows how long?"

"We feel that Daryl's made a lot of good progress, actually."

I took the reins and talked about all of the positive steps he had taken since admission. I talked about how his mood had brightened and he had opened up to me on a number of topics. We had worked on coping strategies that he could take to his new outpatient therapist and build upon in the coming months. Deborah beamed at me while her ex-husband looked like I was trying to sell him a used car.

Then Quadratto interrupted me and told Daryl to go hang out in the dayroom for a few minutes while we took care of some paperwork. After the door closed behind him, Quadratto turned to me and motioned with his eyes to tell them about the 51A.

"Today Daryl told me that he wasn't entirely comfortable living with you," I said to Daryl's father.

"Comfortable? He doesn't get the royal treatment like he does at her place? Yeah, okay."

"He said that sometimes you strike him." The air in my lungs became uncomfortably heavy. "He said you use a belt."

"What about it? You probably don't even have kids. What do you know about discipline?"

"Jesus," Deborah said, shaking her head with a hand shielding her forehead.

"Enough with this shit. Where's the paperwork? Let's go."

"As mandated reporters for the Commonwealth, we were required by law to submit a filing to the Department of Children and Families."

"You did what?! You little piece of shit."

He was turning red with rage, his fists clenched.

"If this is your first reported incident, it's highly likely DCF will screen it out, but we strongly recommend cooperating with them and refraining from hitting your son in the future."

I surprised myself with how steady my voice was.

"This is bullshit."

He turned to Deborah.

"You know it's bullshit. You're just happy because these quacks are taking your side. I'm sure as fuck not paying the bill for this admission, and if you file with DC whatever—"

"We already did," said Quadratto.

"Then you'll be hearing from my lawyer."

He stood up and bolted from the room.

Deborah was shaken up and apologized for his behavior.

"I didn't know about the belt, but I'm not surprised," she said. "Maybe he'll cool it for a while, at least until I can petition for full custody again."

She stood, regathering her composure, and extended a hand to us both.

"Thank you. Oh, and don't worry—I don't think he has a lawyer, unless his divorce lawyer is doing some medical-malpractice moonlighting on the side."

"Take care, Deborah," I said.

"You too, Dr. Stern. Dr. Q, thank you for all of your help, too."

I watched through the glass panel in the door as the three of them walked out of the unit together.

"Well, you did it," Quadratto said. "How are you feeling?"

"Unsettled," I replied.

"Yeah. That never quite goes away."

Rachel and I had begun spending most nights together as we finished off the second year. There had been so much hesitation and so many missed opportunities in the months leading up to our first kiss that we were making up for lost time. We also both knew that we were entering what would be our last year as co-residents in the same program. She would be applying to child and adolescent psychiatry programs around the country, and I would stay on at Harvard Longwood for another year. Anticipating that transition was making me anxious, and it was showing up in our interactions together.

"What's with you this morning?" Rachel asked as we got off the train near the hospital.

"Me? Nothing," I said, deflecting.

We walked in awkward silence, not knowing that Miranda was a hundred feet behind us and had seen us emerge from the train together.

"When are we going to tell our friends about us?" I asked.

"Never," she said, deadpanning.

"Seriously. It's getting ridiculous. We spend almost every night together."

"Yeah, but what if it doesn't work out? Then everyone will know our business."

"So what?"

"I don't want everyone thinking about that every time they see me."

I thought of Erin's recent decision to keep her own romantic foibles to herself. Maybe there was something unique that professional women have to endure, which men are shielded from in this regard. Maybe it was a kind of privilege that I didn't care in the least if anyone knew I had dated Rachel. Or maybe it was just that she was ashamed of being with me.

"What about just our close friends? Just Miranda and Erin."

She shook her head.

"Not yet. If this works out, and we end up together, we'll tell them."

"You want us to out of the blue one day tell them that we're engaged?"

"What's wrong with that?"

"Just seems kind of—"

"Classy?"

"I was going to say crazy."

"Psychiatrists aren't supposed to use that word."

"I think it's okay when used on each other," I replied. "Speaking of which, I'm supposed to see Miranda tonight to work on the M&M conference."

Rachel made a pained face on my behalf. *M&M* stood for "morbidity and mortality." It was a ritual in medicine to have the entire department meet monthly to openly discuss difficult cases with negative outcomes including but not limited to death. The idea is that by dissecting the most challenging treatment courses, we might find learning opportunities from which all participants could benefit. In fields like surgery, M&Ms often address technical approach and concrete mistakes that were made. In psychiatry, it more often feels like a eulogy in search of a deceased patient's character—what defined them as people and how we as psychiatrists could have helped them achieve the kind of life they wanted. In cases that featured bad outcomes, what could we have done differently? The upcoming M&M was to be about Jane. I was dreading it, and Miranda, who had also crossed paths with her on 4 South, had agreed to assist.

When we got together at her apartment, we had a hard time getting started. Miranda is such a bubbly, social person that it can be hard to *not* gab a little. I bet it came in handy with a lot of her closed-lipped patients.

"So what's with you and Rachel? You're like dating now?"

She caught me flat-footed.

"Um, what?"

"I saw you guys get off the train together this morning."

I sat silently, hoping something inspired would emerge from my mouth, but nothing did.

"We're friends," I said.

It was true and not the whole story, which Miranda knew already.

"Friends our age don't have sleepovers."

There was another awkward pause as I scanned the depths of my mind for what to say. I knew Rachel didn't want anyone to know, but I couldn't very well lie to one of my best friends.

"You'll just have to ask Rachel about that," I finally said.

She smiled knowingly.

"Fair enough."

Finally, we got to work on a PowerPoint presentation that coldly went through Jane's life and treatment course. The talk ranged from psychopharmacologic interventions to family dynamics and everything in between. What was lost was the human element of our interactions. The empathy I felt for her and my yearning for her to be well couldn't be captured in a PowerPoint format.

When it came time later in the week to present the case at the M&M, I was amazed at how warm and curiously engaged the audience was. The room was filled with psychiatrists from the department, fifty or so, many of whom had been practicing for decades. One senior member of our community was a direct disciple of Freud's inner circle, or so his reputation held. Nobody cared very much about what medications the patient was on or what court orders were in place at the time of Jane's death. Instead, the group asked me to tell them about her and our interactions. I recounted my experience of her as a young woman who was terrifyingly bright and savvy but so overcome by fear that she couldn't even help herself when given the opportunity. I conveyed the powerlessness I felt in those interactions and wondered if perhaps that was how Jane felt in much of her life as well. In the end, I concluded that I felt privileged to have had the chance to get to know her, and that she had taught me a lot. I wished I could have done more for her. Dr. Redding replied that she felt confident I had done quite a bit.

"Sometimes we do everything right, and bad outcomes still haunt us," she concluded.

Rachel: Miranda thinks you're a huge sketchball b/c of your weird conversation the other day

Rachel: try to stop being a weirdo

me: yeah, well now you know why I was freaking out about telling people. It's going to come up

me: so just to make sure I understand, she was like, "Adam was being weird." And you were like, "I have no idea what goes on w/ him." or something like that, right?

Rachel: she said that she asked you about dating

Rachel: and that you said something vague and told her you'd tell her later

Rachel: i said i didn't know anything, that you had mentioned a few girls from online

me: blegh ok

Part 3

Years Three and Four

28

Don't Just Do Something, Sit There

Psychiatric training teaches a foundational theory of human development that focuses on the emotional milestones individuals reach over the life span. These eight stages, identified by psychologist Erik Erikson, begin at infancy and occur one after another into elderhood. The phases of psychiatric residency seem to follow these stages of development strikingly closely but over the course of just four years. An incoming intern is almost entirely dependent upon the program and senior residents, just as an infant is upon its parents, to get them through to the next stage. Over time, that intern grows and develops and learns skills that inevitably lead to struggles of autonomy versus shame and doubt. A junior resident wants to think she can do everything on her own, but when things go wrong — as they so often do — it hits particularly hard. She may struggle with feelings of low self-worth. In the middle portions of a psychiatric residency, a resident may try to be productive and find his own sense of value. In an ideal program he will find acceptance, encouragement, and even admiration among both his peers and mentors. On occasion, as had happened for Rachel and me, residents even find intimacy, versus the isolation that exists in the vacuum of academic space. Eventually, as time marches on and it becomes evident that an end is near, a resident may begin to wonder about producing a legacy and hope to achieve a career of integrity.

By our third year, I was somewhere in the midst of these developmental stages and just beginning to feel like I belonged. I had finally reached the phase of training where I was able to spend most of my time seeing patients in an office setting who were voluntarily coming in for help. I was handed down a caseload of both psychotherapy patients and psychopharm patients as well as a select few for whom I'd serve in the more traditional combined role of integrated psychiatrist — a singular doctor who incorporates the medical and the psychological with frequent extended therapeutic sessions.

My caseload was diverse. I had patients who were entirely new to treatment — some had just been referred to the clinic by their primary care doctors, and I was the first psychiatrist they'd ever met — and some who had been in the clinic for years, their care being handed down from one resident to the next. Many of those patients went on to tell me that it felt like they were almost faculty at Harvard themselves, teaching the residents how to be better doctors, and this role made them feel useful and proud.

I was also able to see parts of the psychiatric spectrum I had only read about during my first two years of training. I was assigned to care for patients who were just *stable.* They had been on the same medication for years, and they just needed to come in for status checks every several months to get refills. Many had their own separate therapists whose scope better covered the complex intricacies of their inner life. I had never before dared to imagine the calming, otherworldly experience in sessions with these kinds of patients in which I could simply help by doing essentially nothing. Usually, I found that if I tried to make a med change for these stable patients — to simplify a regimen or switch to a different, less sedating antidepressant, for example — I would almost invariably make things worse. Their condition would become exacerbated, or they would develop new, different side effects. If, as one of my supervisors advised me, I engaged in the attitude of *don't just do something, sit there!* I could actually be an accomplice in this patient's ongoing stability and high quality of life instead of a disruptor.

On the other end of the spectrum were patients who were so newly diagnosed that I finally had the experience of being aligned with them from the very beginning. Unlike the many patients who had come to me on 4 South, many of whom arrived with years or even decades of hard-earned notions of what psychiatry could and couldn't do for them, some of my new patients had never met a psychiatrist before. I took pride in the ability to show them that seeking mental health care could be a pleasant, if not always transformative, experience.

"I was nervous to come here, but you don't seem so scary," I was told by one young woman who had been referred by her primary care doctor for panic attacks.

"You don't even have a couch in here. You must not be a very good shrink," stated an older man with attentional complaints.

For these new patients, I was able to truly appreciate the shared sense that together we could try to improve their lives using the knowledge I had accumulated to that point. When I inevitably ran into trouble with patients whose condition was worsening despite my best efforts, I had plenty of supervisors and mentors to run to for help. Finally, the "Never worry alone" slogan that cascaded down the generations of residents coming through our program began to feel like reliable advice. By the time I was a third-year resident with a full outpatient caseload, I could hardly turn a corner of the twisting and winding clinic without bumping into a renowned psychiatrist ready to lend an ear or offer some advice.

I needed that tutelage desperately when I began working with Oren in my Tuesday-afternoon psychopharmacology clinic. Oren was a man in his late sixties, originally from Israel, who had lived in Boston for the last thirty years working as a dishwasher. He described it as a good job for him because he felt productive and useful without the burden of interacting with too many people. He told me that in a typical shift, he could sometimes make it into the restaurant, do his entire workload, and leave without ever saying a word to anyone. This form of work suited him because Oren was a severely paranoid individual. He did not fit the traditional profile of having paranoid schizophrenia, which often arises in early adulthood with a pronounced psychotic break. Instead, Oren found that his thoughts gradually became more clouded and fearful when he arrived in America without fully understanding the culture here.

"It was hard for me," he said. "I had to come for good reasons —I needed to get away from my family—but when I got here, I didn't know what anyone was talking about half the time. If someone made a joke, I began to think it was at my expense. If someone offered to lend me a hand, I thought maybe it was to pull me up so they could push me down."

Over time, his wariness turned to more frank paranoia, eventually leading to an involuntary hospitalization more than twenty years earlier that he said was so traumatic, it would scar him emotionally for the rest of his life.

"I would slit my own throat before I would ever go back to a mental ward," he told me repeatedly at our first session.

"It must have been a truly awful experience for you to feel that way."

He simply shook his head.

"Can you tell me about it?"

"I cannot," he said.

He began to sob loudly.

It lasted for several minutes, at the end of which he apologized. I looked at him quizzically.

"If you can't express your pain in here, with me, then where can you?"

"Thank you, Dr. Stern. I appreciate you."

We were off to a good start, and I arrogantly thought I was in good shape with him. He was technically a patient in my psychopharm clinic, which was designed to be a medication-management clinic with brief appointments, which for many patients were infrequent. Upon starting my third-year rotations, though, I discovered that many patients in a psychopharm clinic treated their sessions like mini-therapy. Even after explaining about what I thought the frame was, patients would usually adopt whatever kind of dynamic they felt like they needed, and I wasn't too rigid about it. Oren would come once every four weeks and talk to me for twenty-five minutes straight. I got the sense that I was literally the only person in the world he could talk to.

I thought that he would benefit tremendously from a low-dose antipsychotic medication to help keep his paranoia at bay.

I imagined his fear and social anxiety fading away and even envisioned him making friends or meeting a romantic partner out in the world he had been hiding from for decades. He adamantly refused the prescription, though, giving no coherent explanation. I suspected that as he had been involuntarily committed in the past, he may have been forced to take medications, which could have traumatized him. The idea of starting a medication or returning to the inpatient service overwhelmed him with fear.

Early in our time working together, I made the offer for meds using different phrases and tones each time, and he rebuffed every attempt. He simply liked to come in to see me every month to use the session like a brief psychotherapy visit. The shortened time frame of the visits and the infrequency of the appointments did not bother him — in fact, it rather suited him because he hated interacting with people and had a difficult time even when it was with someone well-meaning like myself.

I brought up the issue with one of my supervisors, Mark McQueen.

"Got a minute?" I asked as I approached his open office door.

"Oh, sure. Come on in!"

Dr. McQueen was a beloved character in the residency for his kindness and warmth to all comers, but what I appreciated most about him was his undeniable normalcy. Psychiatrists are not naïve about the notion that many who choose to go into our field are a bit eccentric themselves. McQueen was perhaps the exception that proved the rule. He was a late-middle-aged guy who dressed casually and had an office furnished comfortably from a local mom-and-pop shop down the road. He chatted with residents in the hallways about their lives and interests, and in much the same way, behind closed doors he always found ways to ask patients about their actual lives and what was important to them. *Who are the important people in your life? How do you like to spend your time? What do you like to do for fun?* His approach conveyed that as much, or more, could be accomplished by empathizing with the core of a person than by trying to solve the riddles of their psychopathology.

"I have a psychopharm patient in my Tuesday clinic. He's quietly psychotic, I think, but he's doing okay. He's held down the same job for years. He's independent. He doesn't hurt anybody. He just doesn't want any medications, though."

"So what's the problem?" McQueen asked, earnestly perplexed.

"Well, it's a psychopharm clinic," I replied.

"It's a clinic. You're a doctor. He's a patient looking for something that you can offer him. Seems pretty easy to me."

"So I shouldn't discharge him from the clinic?"

"Are you kidding? What for?"

"Maybe to open up a slot for another patient who I could actually help."

"Adam, you *are* helping by having this guy come see you and have these sessions every month. I know it may not seem like it, but your acceptance of him just as he is may be the most therapeutic part of his entire treatment plan."

I thought of Jane. In the months since her death, I had become more willing to accept that maybe I hadn't exactly failed Jane, but I still couldn't believe that I had helped her as much as I could have.

"Can I still keep trying to get him on board for a medication?" I asked.

"Sure, but the alliance comes first," McQueen said. "In fact, you'll probably have a much better shot at getting him to take the pills if first you show him that you aren't just some pill pusher."

Over the next year and a half I met with Oren regularly every month. He never missed an appointment and always thanked me profusely at the end. I got the sense that I was the only person in his entire life whom he trusted. About a year into our treatment, though, the dynamic of our meetings changed when his paranoia began to worsen.

"I take the subway to get to work, and I've begun to notice two men following me," he said solemnly.

"You feel threatened by these men?"

He nodded.

"They work for the train company or maybe they don't. I'm not sure. They wear uniforms, but the badges don't look right."

"And they follow you?"

He nodded.

"They're the same men every time?"

"Yes. Well, no. Sometimes they are different people."

"Do they ever approach you or interact with you in any way?"

He shook his head.

An earlier version of me in residency would have made the mistake of confronting his delusions as such, but I knew that would backfire. If I attacked them from the side, obliquely, by empathizing with his own experience, maybe I could make just a little bit of progress in quieting the alarm bells blaring in his mind.

"It sounds quite frightening," I said. "I do find it encouraging, though, that they don't seem to accost you or really directly interact with you. Do you feel like they're dangerous?"

He shook his head, gazing at the carpeted floor.

"Not yet."

Our session was running short on time. I had to convince myself that he was safe to leave my office, and it was time to try prescribing a medication again. I had earned his trust over the course of months. Maybe he would go for it this time.

"Does it ever seem like you might act on your fears?"

"How do you mean?"

"With violence, even coming from a place of self-defense?"

He was taken aback.

"I would never attack someone. Even in self-defense. I could never even hurt a fly! You know this. You know me."

"I do."

I sighed, and now I was staring at the floor, too, trying to think of how to frame my next points so that they might be received well.

"Oren," I started. "I want you to know a few things before we finish up here today. First, know that I care about you and I know you are a good person. Next, know that you can always page me no matter what is going on, and I will call you back. If you ever feel frightened, scared, overwhelmed, you can also just show up at the emergency room and one of my colleagues will call me to tell me you are there. They'll keep you safe there if you ever feel unsafe. Do you understand?"

"Yes, but—"

"But what?"

"If I go to the emergency room, they will commit me, and I

have told you many times I would rather slit my throat than go to that place again."

"They will not commit you without speaking to me first."

"But, Dr. Stern, what if you told them I needed to be committed. That might be more painful than I could bear."

"I would only ever recommend that if I thought it was the only way to keep you and people around you safe. You have my word."

He grimaced. He seemed to know that what I had just said was not enough to guarantee that I wouldn't commit him if he became frankly psychotic, and he was absolutely right.

"The last thing I want to say before we break is that I still think a medication might be able to help you."

He shook his head, and now his hands were rubbing his thighs repeatedly in a nervous fashion.

"You're clearly very upset by these men on the train," I said, my voice rising slightly.

He nodded.

"I'm offering a medication that could make you feel less afraid. It's safe. It's effective. I wouldn't offer it to you unless I thought it would really help."

"I'm sorry, Dr. Stern. The answer is still no. Thank you again. I will see you in a month."

He left the office in a hurry, and I wondered if the next time I saw him would be in the emergency department or, worse still, on the local news. I didn't think he would become violent toward others or himself, but I couldn't know for sure.

For the first time in more than a year, he no-showed to our next appointment. I called and left him three messages, pleading with him to call me back. I only wanted to know he was safe.

I checked in again with Dr. McQueen and also Dr. Mook, who both advised me to begin to send a letter urging him to come to our next scheduled appointment. The letter would also be entered as part of the medical record, documenting my attempt to reconnect with him. The goal was to get him reengaged, but it was also self-protective, to show that I remained active in his care even after he stopped participating. Regular appointments were essential to his ongoing care, the letter would convey, and if he did not come to our next appointment, his treatment would be terminated. It made me uncomfortable to threaten ending his care, but it was also necessary. I could not keep a slot booked for him if he wasn't going to come to sessions, and I couldn't allow myself to be responsible for his treatment if he wasn't coming to see me.

I waited nervously, carefully watching the clock tick the next time he was scheduled to come in. Right on the hour, he checked in.

"I received your correspondence," he said. "I am sorry. I do apologize. I am not reliable."

"That hasn't been my experience with you, Oren. I'm very glad you're here. I was worried something about our last meeting didn't sit well with you."

"Oh no. No no. You are a very good doctor. I know that. It is my fault."

"It's no one's fault, I'd just—"

"Dr. Stern," he cut me off.

It was the first time he had ever done that. I looked up and noticed for the first time that his appearance had deteriorated dramatically in the last several months. His clothes were in tatters. He had large bags under his eyes.

"I need something from you."

"What's that?"

I hoped desperately that he would ask for a prescription.

"I need a letter that says I am okay."

"What do you mean?"

"I need a letter that says I am not a danger to anyone, that I do not have a mental illness. I need this to carry with me."

"For what purpose, Oren?"

"I will carry the letter and then if anything happens with these men on the train, the police will know that a respected doctor had said I was a good man."

"If anything happens?"

He nodded. I glared at him with wide eyes. *Please, Oren, give me something I can use to avoid committing you right now.*

As if reading my mind, he changed his tone.

"I wouldn't hurt a fly. You know that, Dr. Stern. You know me. I wouldn't hurt a fly. Not even in self-defense. I just need that letter. That is all."

My mind scanned through all of the possible outcomes of our session. I could not give him the letter he was asking for. He *did* have a mental illness. Maybe I could give him a meaningless letter written in medical jargon stating nothing of any substance that might appease him and keep him coming to see me, but to what end? Maybe the time had finally come for me to be totally straight with him, and maybe he would listen to me.

"I can't give you the letter you are asking for because I do believe that you are affected by symptoms of paranoia. It does not make you a bad person or a dangerous person, Oren. You are a good person with a condition that I think might respond to a medication. Why don't I write you a prescription and a letter that will state that you are in treatment and under my care? Those kinds of letters go a long way."

There was a long silence.

I continued: "The medication could be—"

"Thank you, Dr. Stern. No. No medication. Thank you. I am okay."

He stood up abruptly and walked toward me. I never for an instant feared for my safety. He *was* a good person, and I did not believe he would intentionally threaten anyone. Still, I feared that his paranoia might lead him to lash out in what he would perceive as self-defense.

He reached forward and grabbed the doorknob behind me. It closed softly and again my mind raced. I ran to McQueen's office, which happened to be open, and told him what had happened.

"Maybe I should commit him. What if he goes out there and murders two guys on the train?"

"Do you think he will?"

"I do not. He has no history of violence. None. He's been keeping himself safe for twenty years. He has no access to guns. He's employed and has no prior history of violence. His thoughts are ego-dystonic—he wishes he didn't have them. He would kill himself before he killed anyone else."

"Do you think he would do that?" McQueen asked.

"I just don't know. I don't think so. And if I sent him in, he'd be safe for, what? Three days? And then he'd probably be discharged and we'd be right back where we started, but he wouldn't ever trust another psychiatrist again."

"It sounds like you've arrived at an answer."

"So I should do nothing?"

Don't just do something, sit there! I recalled.

"So I should do nothing," I repeated sternly.

His pager went off, indicating his next patient had arrived.

"Thanks, Dr. M," I said, leaving his office in a haze.

For the next two months I scoured the *Boston Globe* every day looking for any sign of Oren. There was none. I had him on my schedule until the end of my residency, but I never saw him again. I hoped that our alliance was at least strong enough that someday he might give another psychiatrist a chance, but I wasn't sure.

My leafing through the *Globe* became less frequent, and over time I finally came to terms with the idea that no matter what he was up to out in the world, our treatment was truly finished.

Had I helped him? For almost two years, I'm sure that I did. After that period is a black box to me, and I will never stop wondering how Oren is doing. A part of me knows that it's unlikely I'll ever find out, and hopes that I won't, for fear of a tragic outcome.

29

Being Present

I sat in the lovely office of the program director, Dr. Redding. My eyes were again drawn to the large windows, and I imagined that someday I could aspire to such luxuries. We were meeting for my annual performance review.

"Adam, this should be pretty easy. Everyone seems to think you're doing a good job."

"What? Really?"

She nodded.

"That's no reason to get complacent. You're on track. Keep progressing and you'll do fine. There are always things to work on."

"I understand."

Just as she had shown on our first day of orientation, Dr. Redding had a remarkable ability to command a room with very few words. She had a stern quality that I hadn't often seen past, but on this day she seemed to relax with me.

"So, is there anything on your mind?"

"Uh, no, I don't think so. I'm relieved to hear you think I'm on track."

"How are things going outside of residency?"

"Fine."

"You know, I hear things," she said with a hint of a smile.

Was she referring to my secret relationship with Rachel?

"You do?"

"It's my job to."

"I see. Anything you'd like to share with me?"

"I don't see any reason to. What about you?"

"No, no. I don't. Thanks, Dr. Redding."

I stood up to leave, but something gnawed at me, and I turned back toward her.

"Dr. Redding?"

"Hm?"

"Let's say—hypothetically of course—that two residency classmates were to find themselves romantically involved. Would that constitute some kind of workplace violation?"

She burst out laughing.

"Hardly a year goes by in this place that two residents don't get together. Sometimes I wonder if this program doubles as a matchmaking service."

I smiled.

"That's a relief—hypothetically, of course."

"Of course. In these hypothetical situations, do keep yourself out of anything involving power differentials though. Don't date anyone you have to evaluate or who has to evaluate you. Got it? That's where people like me and HR inevitably get involved."

I nodded.

"Thanks, Dr. Redding."

In the third year of residency, we were allowed by the program to apply to the state medical board to become credentialed to practice medicine independently. We still had more than a year of training remaining, but our program granted us the ability to moonlight at neighboring hospitals, which was a welcome opportunity to earn extra money. Academic hospitals run almost entirely on the grit and elbow grease of their trainees often working eighty-hour weeks for what amounts to near minimum wage. Moonlighting outside the hospital taught me two very valuable lessons. First, I learned that my skill set was actually quite highly valued by society, and I could be paid very well for the work I was doing. If I moonlighted twice in a month, I was able to effectively double my salary. I also became aware that our program had actually taught me to be a surprisingly competent psychiatrist who could manage whatever challenges walked through the door. Baptized in the depths of the emergency room bunker and up on 4 South, I had actually seen and, to my own disbelief, become adept at managing the very wide range of psychiatric emergencies that tended to occur at outside hospitals. The individuals of our Golden Class also made an explicit point of being available to one another 24/7 for curbside supervision if ever any of us began to feel overwhelmed or uncertain. A quick phone call to Rachel, Erin, or Miranda always left me reassured.

All of that apparent competence didn't arrive without a few hiccups. On my very first moonlighting shift at a stand-alone psych hospital, I admitted a new patient with auditory hallucinations and paranoia. It was a straightforward admission, and I started him on a low dose of an antipsychotic medication called olanzapine. I wrote the order for it on an order sheet at the nurses' station, which felt strange since every other medical order I had written in my life had been through a computer. Still,

I wrote for a single 2.5 milligram tablet to be taken by mouth before the patient went to bed. Three hours later, I was paged by a panicked nurse who realized that there was a decimal point in that order.

"I gave him too much medication," she said flatly.

"How much did you give him?"

"Twenty-five."

"Twenty-five?! That's enough to sedate him through next Tuesday, assuming he ever wakes up!"

My mind raced through the newspaper headlines and court cases that I imagined were imminently coming my way. I would lose my license for sure.

"I'm coming right now."

"What are you going to do?"

"I'm going to sit with him all night and check on him every ten seconds to make sure he's still breathing."

When I arrived at his room, the patient was completely obtunded in his bed. I approached eagerly and announced my presence at the bedside.

"Mr. Jacoby. It's Dr. Stern," I shouted.

There was no response.

"Mr. Jacoby!"

I gripped his hand in mine and squeezed. He reflexively squeezed back, but his eyes remained closed.

"Mr. Jacoby, I'm going to perform a slight sternal rub to perk you up a little bit."

This was a maneuver that sounded more pleasant than it was, involving digging a fist into someone's chest to elicit a response. If he was incapable of responding to that, I'd have to send him to the local emergency room.

I felt around on his chest and found the center of his sternum. I made a fist and placed it in the dead center gently.

Here goes. It's just a man's life on the line.

I began to apply pressure and then more force and then began digging in my knuckles in a semicircular motion.

Mr. Jacoby's eyes popped open wide.

"What in God's name are you doing to me?!"

I exhaled.

"I'm sorry, sir. I just needed to—"

He waved me off with a hand gesture and closed his eyes again.

I stepped a few feet away and out into the hallway. I texted Erin.

me: Anything to do for an olanzapine overdose?

Erin: Any dystonia or signs of dyskinesia?

me: No. Just sedation.

Erin: I'd get an EKG and just monitor. Make sure QTc isn't elevated.

me: On it. Thanks.

I wrote the order for an electrocardiogram and said a little prayer that it would be normal. Mr. Jacoby snored through the entire test, but his QTc was 442ms—entirely normal.

"Now what?" the nurse asked.

"You can head back out. I'm just going to sit with him."

"All night?"

"Just until either he perks up or he doesn't."

I pulled over a chair and leaned my head back. As often as I could remember to, I placed a hand lightly on his chest to make sure he was still breathing. Every ten minutes or so I took his pulse.

By sunrise, I was stirred from a daze when the man suddenly sat up in his bed.

"What the hell did you give me last night? I had a dream you were digging your hand into my chest and giving me all sorts of tests."

"I'm sorry about that. You were given a higher dose of medication than we had talked about. It was our mistake, and we had to make sure you were okay."

"I feel pretty good now actually. Just tired," he replied.

"That's great."

"You look like shit though."

I nodded in agreement.

"Can I go back to bed now?"

"Sure thing. Get some rest."

As my shift ended, I walked to the parking garage exhausted. My first thought was that I was never going to moonlight again. My next thought was to check the schedule to see when I could get back on it.

Our residency overnight call schedule also lightened up in our third year, so I picked up the slack and began moonlighting almost every week. The gig became easier with each attempt.

Rachel and I still weren't public about our relationship, but I had never felt this way before. I had been in love, but never at a time that was reasonable and appropriate to consider marriage. I was entertaining the notion that she was the woman I wanted to spend my life with, and if there was even a chance of this, I needed to start saving up for a ring. I would wonder if I was delusional. After all, she hadn't even told our closest friends about us. I still suspected that she was ashamed of being with me, but then on some quiet nights long after we should have fallen asleep, I would stroke her palm while she waxed poetic about what we would name our first child.

Despite the very real possibility that she would be moving across the country in the next year, I wanted to start saving up for an engagement ring, and moonlighting was the only way to do that.

I was on one of these per diem shifts at a nearby community hospital when I shared the elevator with someone who looked familiar. She was a late-middle-aged woman, and she was squinting at me, presumably trying as hard as I was to figure out how we knew each other. Several seconds passed.

"You're the hospital shrink!"

"Hmm?"

"You're the kid shrink who saw my husband after he came out of surgery."

I smiled.

"You're Charlie's wife."

It had been months since his diagnosis. I was glad to be reminded of him.

"How are you both doing?"

"Well, we're here," she said, motioning to our surroundings in the emergency department. "He needs to get that belly drained again."

I nodded, knowing that liver cancer can leak malignant fluid into the abdomen in the later stages of the disease.

"He'd love to see you. We're in bay eighteen. Come."

"I really shouldn't, but please give him my best."

"Come," she repeated more firmly.

She took me by the hand as the elevator door opened and led me straight to her husband's tiny curtained area of the emergency department.

"Look who I found," she said enthusiastically, pulling the curtain back.

A team of two clinicians were all gowned up from head to toe, together guiding a gigantic needle straight into his protuberant belly. The man on the receiving end bore almost no resemblance to the Charlie I had met months earlier. For a moment I wondered if I had embarrassingly conflated two different patients named Charlie in my memory.

"Hey, don't you knock?!" he shouted back.

It was only after his response that I saw the man I knew. His skin was stretched around the belly but hung loose over his joints, arms, and face. His eyes and skin had a yellow hue, and the paper-thin skin around his eyes and nose was infiltrated by dozens of tiny black veins. But under it all was Charlie.

"Look, Charlie. Look who I have here," she said, motioning to me again.

"Son of a gun. The shrink's apprentice!" he shouted, now with the needle halfway into his enormous gut. "How the hell are ya, kid?"

"Oh, I'm fine."

"You look good," he said as the clear yellow fluid began to filter out of the giant needle. "You've even got a little psychiatrist beard now!"

"Yeah, took a while for that one to grow in properly."

"I haven't fared as well."

I could tell he wanted me to pull up a chair, but I had work to do that was piling up every minute I spent chatting.

"Charlie, I—"

"Come sit. Distract me from this fuckin' thing," he said, motioning to his deflating belly.

"That's a lot of fluid," I said.

"This ain't nothin'. The over-under on this is, what? Seventeen pounds I think. Last time my belly was twice as big and they had to get a second container to hold it all! Twenty-seven pounds. Guy wouldn't tell me if it was a record, the prick."

"You holding up okay?" I asked.

I already knew how he was holding up. His body was breaking down day after day, and he didn't have much time left. His mind was doing whatever it could just to keep the dark thoughts away.

"Ya know, kid, it's been a real shit storm since we last talked. Honest to goodness, I wouldn't wish it on anybody."

He paused to consider his next thoughts.

"I'm glad I didn't off myself, though. That wouldn'ta been right."

"I'm glad, too," I said.

It was the kind of straightforward, plainly honest reply that I would have been embarrassed by in my earlier years as a resident. I would have thought that a real psychiatrist should have something more clever to say. I had learned, though, that a short and true response, raw as it may be, always landed better with patients than whatever my impression of what a guy from Harvard would say.

"And maybe it'll all be worth it, ya know? My doc is getting me lined up for this clinical trial. Some experimental drug. Could be a miracle. Who knows?"

"I hope so," I replied. "Charlie, I—"

"You gotta go, I know. It's okay. Go."

"I'm going to come by again after I've cleared the board. We'll have more time to talk."

He smiled.

"It may be a while. I'm a few patients behind already."

He waved me off.

"See ya, kid."

I left, feeling guilty but not knowing what I could have done differently. I was being paid to do work. I had to do it, and there was a lot of it. I saw a patient with mania, then a teenager who had run away from home and was brought in by police. I evaluated two suicidal patients and two men my age who were coming down from bad trips on PCP, violent and raging at me. By the time I caught up, I was physically and emotionally spent. The sun was rising and I put one foot in front of the other until I arrived back at bay 18 only to find the custodian mopping up the floor.

I walked back to the makeshift office for moonlighters and plugged into the online census. Charlie had been discharged home three and a half hours earlier. Damn.

It was four months later that I spotted Charlie's name on the board back at my own hospital. He was admitted to the oncology unit, but he had been marked for transition to palliative care. I was working on the psychiatry consult service and had no business visiting him, but I needed to see him. Having left him to go do other work hadn't felt right at the end of our last encounter.

I arrived up at the unit and navigated to the room with his name posted outside the door. I was taken aback as I entered to find an emaciated man with tubes coming out of his nose and chest. His eyes were open, but they wandered constantly and without any purpose. I approached him, but I couldn't seem to engage. I couldn't make contact.

"Hello Charlie," I finally said.

There was no acknowledgment from him. His eyes continued to dart without any clear rhythm or purpose.

"It's Adam Stern. I don't know if you can hear me or understand me."

I looked up at the whiteboard to his left to begin to put some of the clinical bits of information together to form a picture. He had developed metastases in his brain, and they were causing swelling that pressed on his brain stem. There had been no miracle for Charlie in the next experimental drug.

"I just came by to see you—"

I paused, waiting for an epiphany that did not come. The sounds of the medical devices chiming were distracting and irritating. I took his limp hand in mine and felt a slight grip take hold. It may have only been a reflex.

"I want to say that I'm sorry."

I waited, but there was nothing reflected back to me.

"I'll let you rest."

There was nothing left to say. I let go of his hand and placed it gently at his side. I have no idea what compelled me, but I left a little note on the whiteboard next to his bed.

Here with you. —AS

When I saw Charlie's name drop off the census, I headed back to the room in search of some form of closure. The bed was empty and cleaned. There were no more chimes from the machinery. It wouldn't be long before the next patient took his place, but I did see that my note was still there. It was with him until the end.

30

California Nightmares

Back in Boston, some of our classmates were contemplating what good friends Rachel and I must be for me to be joining her on her California interview tour. Miranda quietly stewed, knowing something more was going on but not having hard evidence or our permission to disclose it, even if she had confirmation.

On the other side of the country, Rachel and I arrived in San Francisco and checked into a luxury hotel overlooking the enormous San Francisco–Oakland Bay Bridge. We spent a day exploring the city, walking through the botanical gardens and picnicking nearby. The scene was a far cry from 4 South or any other place on the Longwood Medical campus. We drove up to Muir Woods and wandered beneath towering redwoods, then took a ferry out to Alcatraz. We ate dinner on the pier back in the city and watched wild seals sunbathing in the harbor. The next day we rented a car and drove to Sonoma, where we took a private winery tour and dined outside surrounded by picturesque rolling hills and valleys. Some of my moonlighting savings were definitively depleted there.

By the end of our vacation through paradise, I realized I

hadn't even thought about the hospital or any of my patients for days. It was the first time since starting residency that I felt that weight lifted off my shoulders. Then we drove down to Palo Alto and the Stanford medical school campus, where Rachel shifted back into professional mode to interview for the Child and Adolescent Psychiatry fellowship program. I spent the day wandering coffee shops and campus bookstores feeling anxious about what the trip portended for our relationship. I walked around campus aimlessly for hours in a kind of stupor. It was a beautiful fall day, and it seemed like every couple at Stanford had decided to flaunt their love outside.

If she loved one of these California programs, could we ever be one of these couples, or would that be it for us? There would still be another year before I graduated from residency, and I had done the terrible long-distance thing before when I went off to medical school without Eliana. It was a recipe for failure. The longing, lack of touch, and paranoid jealousy that came with being so far away from each other had made me miserable. Even after she graduated and moved to Syracuse to be with me, our relationship never fully healed from that year apart. I had told myself I would never do long distance again, and now I had the added incentives of a niece and nephew along with their parents and my own parents to keep me anchored to the East Coast. I had a feeling that wherever we landed next would be where we put down roots and started a life. I couldn't imagine that life on the other side of the country. I wanted Rachel to like these programs, but I wanted her to like the ones in Boston just a little more.

When we reconnected at our hotel toward the end of the interview day, I tried to read her face, but it was flat. Rachel didn't emote unless there was a good reason to. I often found her to be deceptively warm because her neutral face was so unrevealing that when she did brighten, it felt like the clouds parting and the sun shining down.

"So?"

"It was fine."

"That's it?"

"It's a solid program. Their fellows seem really happy and seem to do well."

"Can you see yourself coming here?"

"I don't know," she said curtly. "Maybe. Let's eat. I'm starved, and we have to get to the airport."

That interaction was pretty much all I got about Stanford before we took a late-night flight down to Los Angeles, where we'd do the whole thing over again for an interview at UCLA. At the airport, as we waited to board late into the evening, we got a little looser with each other. Rachel spotted a gray hair among a sea of black near my temple. I didn't believe her, so she took a picture with her phone and showed it to me.

"Wow. What should we name him?"

"How about George?" she replied.

"Like King George. One to rule over the rest."

"Wasn't King George the *mad* king?"

"Maybe with years of psychotherapy he can be downgraded to *neurotic* George or *simply narcissistic* George."

"He'll need a real therapist for that, so you're going to have to ask around for a referral."

When we landed in LA, it was like entering an entirely different world. I couldn't get over the never-ending traffic, but Rachel seemed to enjoy gazing out the window at the sprawling city all around us. We spent the next day exploring Santa Monica, taking pictures with our hair blowing in the wind off the pier. We shopped and dined, again pretending to be wealthy. With the money I had been saving up for her ring, I almost felt like a rich person.

The next morning I dropped her off at UCLA Medical Center, which was located in a gleaming area of high-rises. I didn't recognize anything and felt every bit of the three thousand miles we were from home. I gave her a kiss and wished her luck.

"Text me when you're finished, and I'll come get you."

I was intent on spending the day on Venice Beach. It was one of the parts of our trip I had been looking forward to most. I had glowing, fuzzy memories from early childhood visiting my uncle who used to live near there. We would walk along the beach, stopping to admire the street performers. One did an extremely convincing impression of a walking, talking android that I still think about from time to time. In my memory it was a place of wonder and warmth, but when I arrived again two decades later, I was taken aback.

I looked up and down the street facing the beach trying to see hints from my youth, but mostly I just saw pot shops, tattoo parlors, and novelty shirt businesses. It was seedy. There was a visceral grime to everything around me. Even the air seemed gray that day.

I stopped for lunch and headed back to the hotel around midday even though I had paid for all-day parking. The entire scene made me feel a certain kind of sorrow for what was lost. Had it always been like this, and I was the one who had changed?

I waited, anxiously checking my phone every few minutes.

It began to rain — in a place where people only rave about the weather — and a sinking feeling settled into my stomach. *I am not moving to LA,* I thought to myself.

Finally, mercifully, the text came and I jetted down to the car to go pick Rachel up.

When she hopped in, her demeanor was beaming. It was jarring seeing her so excited.

"You liked it?" I asked flatly.

She nodded, smiling.

"It's a really nice place. All of the fellows have their own offices that have like floor-to-ceiling windows with amazing views. They all seem super happy, and they specifically have a research track with guaranteed funding."

The rain poured hard on the windshield.

"Sounds great," I said with feigned enthusiasm.

"What?"

"What? Nothing. I'm glad it was a good program. Should we go back to the hotel?"

I put the car into drive.

"Isn't it a good thing that I liked it so much?"

I nodded.

"So why are you being weird?"

"Why do you think?" I asked, putting the car back into park.

"Because it's in California?"

"Specifically, because you would be in California if you came here."

"Well, yeah."

"I just don't know what that would mean for us."

"Yeah. You would just come here in a year if we were still together, right?"

"I don't know. It's a lot to consider," I said.

On the one hand, it felt good to hear her think in terms of us being together that far down the road, but I also resented that our current relationship wasn't even a consideration in her calculus. I wondered if it should be, or if, rather, I was being unreasonable. I would never want her to choose a program back east for me and then spend the rest of her life with me resenting me for the missed opportunity.

She pulled out her phone.

"Who's texting you?"

"My friend from college. She just got engaged."

"That's nice."

"When do you think we'll get engaged?"

"What?"

"Do you think we'll get engaged?"

I felt my heart start to pound and the blood rush into my cheeks.

"I don't know. We don't know if you're even going to be around next year, and—"

"And what?"

"You won't let us officially become a couple."

"This again?"

I nodded.

"It's not right. It feels like you're ashamed of being with me."

"I'm not ashamed of being with you," she said, putting a hand on the back of my neck.

"I've been saving up for a ring, you know."

"You have?"

"Yeah, but—"

"But what?"

"It seems ridiculous to even think about getting engaged when no one even knows we're together. Maybe we can talk about getting engaged sometime after you let me — I don't know — tell our best friend about us."

"I told you. I don't want everyone at Longwood to know our business. That's my worst nightmare."

"You're about to be gone from Longwood. You'll probably be three thousand miles away next year. What difference does it make if people know your business?"

"And what if I do move three thousand miles away? What then?"

"I don't know," I said angrily.

A moment passed and I gripped her hand tightly in mine. We sat there in the car, listening to the rain and holding hands across the center console. As close as we were to committing our lives to each other, inside it felt like we were on the verge of breaking up — or at least deciding that we were on a path to break up someday if she moved to California for a fellowship. In that case, our relationship, which had only just begun, would already have an expiration date.

Miranda: Hey! question for you

me: hi

Miranda: I just realized that the all-resident's meeting is on Wed

Miranda: do you know where it is?

me: no idea, I just go where they tell me to go on the day-of email reminders

Miranda: haha yeah

Miranda: how has your weekend been?

me: Magoo (my guinea pig) passed away this morning, so it's been a tough day

Miranda: oh no! I'm so sorry to hear that :(

Miranda: yeah — very tough

Miranda: how old was he/she?

me: yeah, she was very old and seems to have gone peacefully

me: 7 years old

Miranda: I'm glad to hear it was peaceful

Miranda: wow — that's a long time; I'm really sorry

Miranda: death sucks

me: yeah. I had a therapist for a bit, but I feel like Magoo was the best therapist. Always listened and never billed me.

Miranda: lol

me: Sounds ridiculous, but she really made me feel like I wasn't alone all the time, I mean when Rachel isn't over or whenever I actually was alone.

me: not to be morbid but then this AM I had to figure out what to do once I had "pronounced" her . . . so I was doing all these google searches for pet cremation, etc. cause I didn't know what was appropriate.

Miranda: :(— I was going to ask if you buried her

me: eventually I took her to this MSPCA animal hospital in Jamaica

Plain where they would cremate her for a reasonable fee and scatter the ashes in their cemetery. they were very nice

Miranda: the best it could be under the circumstances . . .

me: right

Miranda: :(well, I'm really sorry about her death :-/ really sux

Miranda: r u getting together with Rachel?

me: yeah, we were just talking about her coming over here and getting dinner

Miranda: oh nice :)

me: we're thinking legal sea foods, if you're interested in joining

Miranda: up to you — totally understand if you just wanna spend time with her. She told me about you two, btw. Not that it was much a surprise.

me: oh, that's great! One less totally unnecessary secret to walk around with!

Miranda: would be nice to see you guys

Miranda: been awhile with us all on different clinical rotations

Miranda: meeting there at 6:30?

me: yeah, let's plan for that unless you hear from me

Miranda: ahh sweet — maybe I'll walk — and that will be my exercise for the day

Miranda: cool! well, I'll see you guys soon :)

me: nice, see you then

31

Not My Choice

By the time we were thoroughly into our third year, Feelings class had become an environment like none other. Our group of fifteen residents and two attendings had become so bonded by years of intense shared experience that we engaged together much in the way that siblings in large families do. We fell into well-defined roles. Some members of the group consistently railed against the injustices of whatever rotation they happened to be on, ever the familial victims, while others, such as Drew, became an unending source of level-headedness.

He once said, "We're all under such stress carrying the cumulative weight of all of our patients, it almost makes sense that we would feel like no matter where we rotate, we are unable to protect ourselves."

Even Nina and Jen seemed to look to Drew to counter some of the more emotionally tense assertions made within the ultra-safe confines of our group. It became an open joke. When Dana, Ben, or Gwen took a turn calling their attending uncaring or their supervisor out of touch or their patient sociopathic, heads instinctively turned to Drew for calm and rational reassurance.

Sometimes, though, not even the combination of Nina, Jen, and Drew could seem to lower our boil back to a simmer. There was one class when Svetlana talked angrily about how the program would not support her taking time off to align with her eleven-year-old daughter's vacation. Most of the group hadn't even been aware that she had a daughter and were caught too off guard to respond usefully, just as I had been at the bar months earlier. In that Feelings class, though, Erin stepped up, revealing to everyone's surprise that she was pregnant again and couldn't imagine anything—not even residency—ever competing again with her child. Svetlana seemed reassured that someone could identify with her struggle.

"I used to know things about myself," Erin said sadly. "I knew that I had to keep achieving just to prove to myself that I was worthwhile. Maybe if I just got into Honor Society, then my parents would appreciate me. Maybe if I just got into a good college, a good medical school, Harvard for residency. I was on this never-ending trajectory. Even if I became chair of the entire department, I'd still be trying to prove my worth, to show that I have value. It took having this baby inside me to realize that the things that I thought were important to me aren't that important to me now."

"So what *is* important to you now?" Jen asked.

"It all starts here," she said, motioning to her abdomen. "For this child to come into the world she deserves, I need a good marriage, which means I need to improve my relationship with Bobby. Period. She deserves to have two parents who are actively fulfilled in their life, and if that means that I take a job out of residency that's less prestigious so that Bobby can live in a city he doesn't hate and find work that won't kill his soul, then that's what I need to do."

"How do you know that you won't just accommodate Bobby and have him be exactly the same, just in another situation?" I asked.

From everything I knew about Bobby and Erin, they *worked* in a certain way because she bordered on a never-satisfied superhero, so it just didn't matter if he didn't have a job or want to socialize, because his role was to support her in hers. With so much uncertainty and self-doubt, Erin felt like she couldn't thrive without Bobby. Still, I found myself feeling defensive on her behalf. Didn't she deserve to pursue her dreams?

"Wouldn't it be unfair for you to give up what's always been important for you without knowing for sure that it would work out?" I asked.

As the words left my lips, I realized that the same could be said of Rachel's and my relationship if she chose me over a fellowship program or if I chose her over being near family on the East Coast.

"What I'm hearing is that you feel like your ambition needs to take a backseat in order to have a chance at the life you want?" Nina asked.

She shook her head.

"It's the same ambition, but I'm going to apply it to my husband and family from now on."

We all seemed to fall into a kind of ponderous stupor because the topics that were being raised applied to each of us in our own unique way. I guessed that we were all thinking quietly to ourselves about our own paths forward as Nina and Jen talked about our new reality, that each passing decision we made closed a door behind us.

"For so much of your lives, you've gotten from one stage to the next to the next and continued to have the entire world open to you, but now as you begin to think about what lies ahead, you've been confronted with the reality that there are trade-offs," Nina said.

"Like you, Rachel. If you fast-track into child and adolescent psychiatry, you lose that last year of residency and that will never come back around, but what you gain will hopefully be more aligned with what you want out of your career, and out of your life," Jen said.

"The rest of you who will be with us through next year still have those decisions to make, and let me tell you, they don't get easier, not even after you're long out of residency training!" Nina said. "I've been a professional psychiatrist for a long time now, and I'm still never sure that the choices I make in balancing my personal and professional life are the right ones."

I looked across the table at Rachel, who, per usual, was looking down doodling flowers in a notebook. I smiled, finding, as always, that her casual disengagement was charmingly antagonistic. But the decisions we had in front of us could define the rest of our lives, and I needed to know where her head was. Since our California trip, she had also interviewed at Mass General, an excellent Harvard affiliate hospital across town in Boston. It had a spectacular reputation and often took on fellows from our residency program. She liked it a lot, but was it enough to rank above UCLA? If she ranked Mass General first and matched there — two big *ifs* — we'd have a shot, and I would buy the engagement ring the next day. If she ranked another program like UCLA first, or if she just happened to match there despite ranking Mass General higher because of the Match algorithm, the proverbial ball would fall into my court, and I would have to make the decision of sacrificing the vision of my life on the East Coast. Meanwhile, she still wasn't telling anyone beyond our best mutual friend, Miranda, that we were dating. I had to ask myself if she was as committed to our relationship as I was, and if she wasn't, wouldn't it be a fool's decision to follow her to California if that's where her match led?

Our dynamic reminded me of the couple I was working with as a couples' therapist. In residency, as part of our outpatient training, we were each assigned a couple seeking therapy. Terrence and Xavier were my couple, and they reminded me of Rachel and me so much that I had to seek supervision about how to deal with my own biases.

Terrence and Xavier had been together for over a year and were privately very much in love. Terrence came from a conservative Christian background and hadn't told his family about their relationship or even that he was gay. Xavier wanted to get married and adopt a child together, but Terrence wouldn't do it.

"I can't get married without my family there, and I just can't tell them. If I tell them, my momma will disown me. I can't do it."

"So what if she disowns you, if she doesn't know who you really are?" Xavier replied.

"It matters to me."

"It shouldn't. Should it? Dr. Stern? Should it?"

I wanted to come clean that they were my first couple and it seemed like they had real problems that warranted seeing someone more experienced, but I recalled the lessons Dr. Mook had imparted and stuck to the script. I reflected back what I had heard each say and minimized my own feelings that heck yes Terrence should tell his family and it was an affront to Xavier that he hadn't already done so. Eventually, against my therapeutic advice, Xavier gave Terrence an ultimatum that he had to tell his family or their relationship could not continue. Terrence simply could not handle the pressure being placed upon him. The parallels to Rachel and me became overwhelming when Terrence was offered a job out of state and decided to take it.

"So I guess this is it," Xavier said over the phone. "T won't even come to a final session with me to reason it out with you."

"I'm so sorry," I said to Xavier. "I'm here for you both if I can help."

I had kept a solid stoicism in my voice, but internally, I felt like a failure as a psychiatrist, and felt a terrible foreboding of what it would mean for me as a partner, too.

That night I asked Rachel what her rank list was going to be. It was due the following week. She said she didn't know. She was still considering the exact order. I wondered if she truly did not know or if she was simply avoiding the inevitable downfall of our relationship. If she told me she was ranking UCLA first, then I would know where our relationship stood, and it would begin a cascade of events that would eventually lead to our breakup. I was sure of it. Either I would choose to end the relationship or it would become so strained by distance that it would weaken and end, like my relationship with Eliana had years earlier. Whether it was because of Erin or Terrence or any other of a million voices guiding me forward, I decided in that moment that I would absolutely not force her hand. If she wanted to be with me, she would need to make that decision purposefully, without my having pushed for it. If she chose to stay in Boston simply because I implored her to do so, she would resent me forever, an even more detestable outcome than the anticipatory grief I felt for losing grip of our relationship. I had to give her the room to make her own decision.

Rachel: i think you should stay over on tuesday

me: yeah, if you want me to be there Wed for the match, I should stay Tues

me: tues is tomorrow

me: this week is flying by

Rachel: yeah

Rachel: ha

me: i'll come home after work. and let you get home and settle in and then i'll come over. even if you're in a bad mood after your long day, deal?

Rachel: ok

Rachel: hopefully i won't get home at midnight

me: we can touch base at some point and figure out if we want to eat something together

Rachel: ok

Rachel: usually i grab something small before group and then eat like a yogurt or something when i get home

Rachel: so feel free to eat if you want to

me: ok

me: i'm getting started on laundry, back in a bit

Rachel: BYE

me: i just discovered your body butter in the bathroom. did you mean to leave that here or do you want me to bring it tomorrow?

Rachel: no leave it

Rachel: i found one that i like better that i'm using at home

Rachel: so that one is for your house

Rachel: you can use it if you want

me: am i supposed to put it on bagels?

Rachel: not funny

Rachel: but i dare you to do it and eat one

me: i honestly don't know what body butter is for

Rachel: it's like lotion

me: i mean i guess it's to make your skin smooth but i don't know where or how to put it on

Rachel: you put it on wherever you want smooth skin

Rachel: i'm going to shower and get ready for bed

Rachel: i'll see you in the AM

me: ok, gnight

32

The Match Redux

When I arrived at her apartment, her roommate, Julia, let me in. With sweaty palms, a pounding heart, and a bouquet of flowers, I tried to act casual.

"Oh, those are nice! Rachel's just in her room," Julia said. "Come on in."

I walked down the hall to Rachel's room and knocked on the door.

"Come in."

I walked in and presented the bouquet to her. She offered up one of her authentic smiles, the kind that only a lucky few got to see.

"These are pretty."

"They're celebratory flowers," I said.

"Or consolation flowers."

I took her hand.

"However this goes, you deserve to be celebrated."

She smiled again and leaned into me.

"These are going to be a *long* sixteen hours."

I had arrived to be with her so that when the Match results were released online the next morning, we could share in that experience together. Once again, the same mysterious and all-powerful algorithm that had landed us both at Harvard Longwood was tasked with determining where Rachel would do her fellowship training. It took into account both her priority list and the programs' rankings of each applicant.

We were both playing hooky from didactics, but it was late enough in the game that we weren't scared to break a rule on special occasions.

"There's no way I'm going to be there in front of everyone when I find out," she had said. "What if I don't match anywhere?"

Her comment fit perfectly with her need to keep things close, even with me.

"You'll match," I said, reassuring her.

We hardly slept at all that night. We talked about all kinds of things — the nonsensical and the serious and everything in between. When at around 3 a.m. Rachel again suggested a name for our first child, I sighed.

"What?"

"Let's just talk about that in a few years, or at least in the morning."

"It's already morning."

"Fair," I replied. "What do you think of Magoo? Works for a boy or a girl."

"Ugh. I changed my mind. Let's go to bed."

When the sun rose and the birds began to chirp, I got out of bed and walked a few blocks to the coffee shop down the street. I ordered two coffees, including her fancy caramel macchiato, and headed back to the apartment, where I found her wide awake and pacing in her ten-by-ten-foot room. We watched as the clock ticked and ticked and finally noon arrived, and she loaded the website that would tell us our fate.

She hit Submit as I stood three feet away. Then she grimaced.

"What is it?"

"The website crashed."

"Try it again."

"I'm trying."

I came over to her side and looked at the screen. She reentered her elaborate username and password. This time when she hit Submit, the screen read:

Congratulations! You have matched:
Massachusetts General Hospital

"Mass General!" I exclaimed.

I squeezed her in my arms and felt tears begin to fill my eyes. The elation lasted for only a moment when an insecure thought invaded my mind. Maybe she had ranked UCLA first and had matched close to me by accident. I loosened my grip on her and pulled back just enough to look at her face.

"I never asked you what you ranked first," I said.

"This," she replied.

She was genuinely beaming.

I think she meant Mass General, but it could have just as well been *us* that she had chosen. I had known how I felt about her from the time I met her, but it wasn't until that moment that I knew without a doubt how she felt about me.

• • •

Our lives changed very quickly after the match. I began doubling down on moonlighting to save up for the engagement ring, and we decided to move in together when our leases ended that summer. Rachel graduated from our program a year early and entered fellowship in one fell swoop. Just like that, we were living together and working apart. We had never done it that way before, but our relationship thrived. Residency is what had brought us together, but having one work world and a separate domain for our relationship functioned a lot better. The distinction made the relationship feel solid all on its own. She still knew all about where I worked, and all of the people surrounding me, which occasionally made for interesting gossip. I had to learn all about her new fellowship classmates. And our friends and colleagues all finally knew, without question, about our relationship. In fact, we had become "Facebook official," which at that time felt as momentous to me as the engagement I had been meticulously planning for months.

Rachel had waxed poetic enough times about her perfect ring for me to know what I wanted to get her, but I had no idea how to find the right jeweler. My mother assured me that there was an entire world of diamond jewelers that would do a better job designing the perfect ring than if I just walked into the mall store as was my first thought. But my mother only knew about the scene in New York. She took it upon herself to call up our old next-door neighbor, who lived in Boston and was "a man who knows all things class" — whatever that means — and he came through, hooking me up with a jeweler in Boston's diamond district who helped design the perfect three-stone trilogy ring. I wrote the biggest check of my life to that point and carried that ring around for a month, checking for its whereabouts every thirty seconds or so.

I had decided to propose on the roof deck of our building, which overlooked the Boston skyline on one side and the Charles River on the other. I was returning from a conference in New York, having schlepped the ring with me, always in my breast pocket. Upon my return to the apartment, I took champagne and a little music player from my car into the building along with the ring. I taped a little handwritten note that simply requested the pleasure of her company up on the roof and knocked on the door before sprinting up the four stories to the roof deck.

To my utter dismay, I found a man sitting up there. He was a neighbor we knew casually, only enough to occasionally nod hello. It was a shared roof deck for the entire building, and it was a beautiful fall day, so I should have foreseen the conflict, but it just hadn't occurred to me that such a scenario might arise. Flustered, sweating, wearing a full suit and carrying a bottle of champagne and a music player, I approached the man, who was reading a newspaper.

"I'm sorry to interrupt your afternoon, but I need to ask an enormous favor."

"Hmm?" he said, barely looking up.

"I need you to please leave."

"Excuse me?"

"My girlfriend is walking up those stairs. I'm about to propose to her. I'm so sorry, I didn't think anyone would be up here. Do you mind?"

If he said *yes,* he did mind, I had no idea what I would do.

Throw him off the roof? Drag him down the stairs by his ear? No appropriate solutions immediately came to mind.

Thankfully, he agreed.

"Beautiful day for an engagement," he said, standing and stretching slowly.

I stared him down, my eyes urging him to move more quickly toward the stairs.

Little did I know that there was no rush. Rachel was out shopping and wouldn't return to find my note for another hour. I stood up on that roof deck pacing for the better part of the afternoon, wondering if she had figured out what was happening and decided to flee.

Eventually, though, the stairwell door opened again. We came together and instinctively started swaying to the slow music playing from my portable speaker while I gave a rehearsed speech that neither of us can remember very well. We were too swept up in the joy of finally being together, fully. I showed her the ring and asked her to marry me, and she said yes. Standing above the city that introduced us, we kissed for the first time as an engaged couple.

33

I've Got Muscles

The last year in the Harvard Longwood residency program offered a great deal of flexibility. We all spent half our time seeing therapy and med-management patients in the outpatient clinic. What we did with the remaining time was more varied. Research-minded residents such as Drew gravitated toward research projects that would catapult them into academic careers. People who wanted to specialize in particular areas that didn't require a full fellowship after residency spent time pursuing those interests. For example, Svetlana had spent time in the US military prior to medical school and signed up for an elective clinical rotation at the local VA hospital. Miranda and Dana entered into a nearly monthlong cold war, each vying to become the program's next chief resident. There was secret-keeping and an abundance of coy half measures when asked about what they were applying for. It was an unspoken competition that we all knew was happening but could only mention in hushed tones. The competition was so intense and wearing, even on the administrators, that at some point Dr. Redding asked me if I would be applying for the role. She was probably looking for an easy out to avoid entering the war between Miranda and Dana.

I told her that while I was honored to be asked, and would be happy to interview, I had my eyes set upon another chiefship.

After an elaborate interview process, the powers that be chose Miranda, who I knew would make an excellent program chief due in no small part to her never-ending effervescence and optimism. In a program where overworked and underappreciated residents sometimes fell prey to despair, having a bubbly and engaged person at the head was smart, and Miranda went on to do a better job than I ever could have.

The gig that I was shooting for was to be the outpatient chief, managing the resident caseloads and supervision, while reporting directly to Meg Mook. Erin applied for this position, too, and as always in her life, she excelled as an applicant. She was always more prepared and readier to lend a hand than I was, but I had also begun to win Dr. Mook over with my performance in our Tuesday-morning therapy seminar. When the chiefships were announced, the position was split into two slots, with Erin and me assigned to be co-chiefs. We were delighted to be working together again.

With her on board as co-chief, I had some extra time available to pursue another interest that I'd had since watching Deborah's amazing transformation through electroconvulsive therapy. There was an entire field within psychiatry devoted to neuromodulatory interventions, the process of affecting change in the brain through stimulation. Electroconvulsive therapy had been around for decades, but there were newer approaches, too, like transcranial magnetic stimulation (TMS), which our institution just happened to be a world leader in. The approach had only been around as an FDA-approved treatment for about five years, and some of the early pioneers in the technique worked in our neurology department. I had wanted to learn more about it, and this was the perfect opportunity. I decided to rotate on these services again and truly learn to practice psychiatry with brain stimulation.

The targeted approach in the TMS clinical program amazed me. While electroconvulsive therapy was a blunt instrument that caused the entire brain to pulsate with electrical activity, the TMS device used magnetic coils to encourage tiny sections of the brain to activate. I learned that the brain has so many overlapping functions that we evolved to develop functionally connected circuitry — parts of the brain that are interdependent. There are well-defined circuits involved in mood regulation, and TMS had the power to tap into a particular node of depression networks to modulate several parts of it. It was like turning a single screw that was so elaborately connected to the rest of an apparatus that the whole complex system was affected — almost like a Rube Goldberg contraption.

The neurologists who ran the program were renowned clinical neuroscience researchers. They seemed to be driving forward together toward a unified goal of changing the world through good neuroscience. The premise was that we now had the technology to target just the right spot on the brain, and in doing so, we could make all the difference in a patient's outcome. It was a stark contrast to the looser techniques of psychotherapy and psychopharmacology that I had trained in up until that point, and a refreshing change of pace.

I was initially drawn to these neuromodulatory techniques because it felt like a relief that the patient's outcome did not actively depend upon what I said or how I said it. It was a biological intervention without the shackles of interpersonal dynamics. With experience, I learned that I was wrong, of course. The more exposure I gained in this area, the more I came to realize what Dr. Macy had taught me years earlier — the person at the center of care remains the most essential piece of treatment outcomes, and a psychiatrist's ability to connect with that person is likely to be at least as important in determining outcomes as the biological effects of an intervention.

Over the course of that fourth year, I began to master the technical aspects of the treatment, but mainly I learned about how to connect with patients while they were coming into the hospital to have their brains zapped five days a week. There was a new kind of rhythm I had to master to get to know someone new, work with them intensively for a couple of months, and then wish them well and send them back to their referring outpatient psychiatrist. When my patients got better, I took pride in seeing that process unfold in such a short time. When they didn't respond to treatment — almost half the time — I felt dejected but doubly motivated to help them find the next option for their care. More than a few times, I facilitated referrals to Dr. Macy over at ECT when someone hadn't responded to TMS.

Another lesson I learned in that wonderful fourth year was that I loved teaching junior residents and medical students. I found that I was never more engaged at work than when I was trying to explain a complex therapeutic concept to someone hearing about it for the first time. I lamented that medical students had very little exposure to the entire world of outpatient psychiatry, more often being assigned to a treatment environment like 4 South, so I created a pilot program to change that. Rotating medical students joined my psychopharm clinic appointments with a subset of appropriate patients who agreed to have an observer sitting in on their sessions. Afterward, the students and I spent hours talking about all of the pertinent concepts that had come up. I found that there was no better way to keep my own knowledge fresh than to have to explain the important concepts to others.

Dreaming up and implementing that experience was the first time I seriously considered staying in academia after graduating from residency. I had learned in residency that academic institutions such as Harvard considered teaching to be a shared responsibility that did not come with much, if any, additional money or protected time. On top of whatever academics' day jobs were, whether doing clinical care or scholarly research, Harvard Medical School faculty were *expected* to teach trainees. It was generally a responsibility that came with few tangible benefits, monetary or otherwise, but the people who chose to become faculty did so because there's nothing quite like helping to guide the brightest minds forward. To this point, there was no doubt in my mind that Harvard Medical students and residents were, on the whole, brilliant, but I was also struck by how ordinary they seemed at the same time. As I had learned when I arrived, folks at Harvard had as many imperfections and insecurities as anyone else; they also tended to pair those qualities with an amazing ability to soak up and apply knowledge.

The feelings I had throughout much of my residency — that I was inadequate, that I had matched at Harvard by mistake, that I wasn't as good as those around me — were finally beginning to fade in the beginning of the fourth year. The peak of my confidence came when we received back our PRITE scores. PRITE stood for Psychiatry Resident-in-Training Exam, which was taken annually by all psychiatry residents across the country to assess their progress in training. In my first three exams, I had simply not wanted to embarrass myself and studied hard enough to do that. When Dr. Redding told me I was "on track," she based that statement partly on a PRITE score that was rising steadily but still in the middle of the pack. By the fourth-year exam, after moonlighting everywhere around town and practicing independently, I was feeling so good that I didn't feel the need to prove myself to anyone.

I walked to the conference room where exams were being handed out and ran into Tina, the administrative person who had first introduced us to the residency years earlier.

"Hey, Tina."

"Oh hey, Adam."

She was carrying two sealed cardboard boxes with all of the PRITE materials inside.

"Can I give you a hand with those?" I asked.

"I'm okay. I've got muscles," she replied.

"Fair enough," I said.

I dug around my pocket to find a number 2 pencil.

The exam began, and the time breezed by unlike in any of the three prior attempts. I finished with half an hour to go and handed it in to Tina at the front of the room. She smiled politely, and I halfheartedly flexed a bicep.

When the scores came back, I had gotten the highest mark in the entire residency. Some lessons along the way had clearly begun to sink in. Truly, I did belong.

I was engaged to Rachel. I was thriving at work and living in a city that I loved. I was feeling simply untouchable. That's when my new patient Elise walked in the door and changed everything.

34

An Impostor Once More

"Why should I even listen to you? You're not even a full, real doctor yet. You're just the first guy that could see me."

All ninety-two pounds of this twenty-year-old were seething at me from a chair diagonally across my office. Her eyes glared with such intensity that I wanted to look away. I was twice her size and yet felt like I could fit in her pocket. Simply put, to me she was Jane reincarnated.

"You're not wrong," I said. "Well, the last part wasn't entirely accurate. As chief resident on this service, I'm kind of in charge of which psychiatrists patients are assigned to, and I can tell you that a lot of thought went into your pairing with me."

It was true. Dr. Mook knew how traumatized I had been from losing Jane and suggested that it would be good for my training to have another patient with anorexia nervosa to work with in therapy. This patient's chart had been littered with phrases such as "cluster B traits" and "character pathology," which I had learned to be psychiatric code for *personality disordered*. Taking on her case was the last thing I wanted to do, but I respected Dr. Mook's opinion enough that I didn't fight her recommendation.

"So what, you're some kind of hot shit 'cause you're head of the little doctor dweebs?"

I didn't know what to do with all of her vitriol, but my face must have communicated the sting I felt because she relaxed in her chair and looked away. Her eyes landed on my bookshelf.

"*Freud and Beyond*? What is that, like Buzz Lightyear? *To your unconscious, and beyond!*"

"Do you like *Toy Story*?"

"How old do you think I am? Jesus. Are you —" She paused to try to find a room label of some kind. "Isn't this a Harvard-affiliated hospital? God help me."

We sat in deflated silence for a full minute.

"Maybe we can talk some more about what brings you in. You are an adult, so you're here seeing me voluntarily, right? That tells me that a part of you wants to feel better."

"Do you hear yourself, Adam?"

I had gained enough experience to know that when I introduced myself as Dr. Stern and a patient responded with Adam, it was usually a purposeful effort to cut me down. If that was her intent, she was killing it. In that moment, I felt smaller than nothing; I was truly an impostor again.

"Let me explain something to you, Adam. I am here because Boston College won't let me back next semester unless I am stable and in treatment. That is the only reason that I am here."

"So, you do have a goal that I can align with. Let's work together to get you back to school next year."

There was another heavy silence. This time our eyes were locked, and I wasn't backing down.

Finally, without warning and with twenty minutes left in our session, she stood straight up and leered over me.

"Fuck you," she spat as she stormed out the door.

"It's good to see you again," Dr. Pettyjohn said with a warm smile.

Her office hadn't changed much in the months since I'd been in.

"It's nice to see you, too, but truthfully I'd rather not be here."

"I get that a lot," she replied. "Tell me more about what brings you back in."

I brought my old therapist up to speed. I told her about Rachel and our engagement and the feeling of finally getting to work in the outpatient setting, which is what I'd always wanted.

"That all sounds pretty wonderful."

"It does, which makes it all the more painful to realize that I'm in my fourth year of this residency program, about to finish and enter the world as a real psychiatrist, and I don't even know what I'm doing—like at all. I've studied. I read the texts, and I've tried to put them into practice, but most of the time I still feel totally powerless to help. I see the confidence and natural rhythm that my attendings show off on a daily basis, and I just think, *That's not me and that will never be me.*"

I was looking down at my shoes, trying to will the tears to unform.

"What if I really did match at Harvard by some kind of fluke, not a mistake but a random act of universal nonsense, and no matter how hard I've tried, and no matter what I do for the rest of my career, I never feel like the real thing?"

"You feel like a fake?"

I nodded.

"That's a pretty common thing around here. As you probably know, there's even a name for it. *Impostor syndrome,* as though it's an illness and not just something that we all go through on our way to becoming a more self-actualized version of ourselves."

"You went through it too?"

She burst out laughing.

"I was so sure that I was a dud in residency that I came to see someone in this same office for my own therapy."

"Really?"

She nodded.

"And then something funny happened. I started running into my classmates in the waiting room, one by one, until eventually we all started to realize that every last one of us was just trying to get by and figure it out as we went. I'm sure these classmates in your golden whatever have felt just as lost as you do sometimes, too."

I thought of Erin's combined brilliance and insecurity, and all of the cumulative naïveté expressed in our Feelings class. Dr. Pettyjohn was probably right.

"Suppose that you're right," I said. "Suppose that everyone around me struggles with feeling illegitimate as much as I do. Wouldn't that mean that psychiatry itself, as a field, was some kind of a sham?"

It was a question I had been afraid to ask for nearly four years. The field of psychiatry has a well-earned reputation among its medical peers as being pseudoscientific. It's as far from mechanical and physiological as one could get. There was always going to be more subjective assessment than objective data in mental health care. A psychiatrist can't even know if a patient's mood is better without asking him.

"What do you think?" she asked, volleying the question back to me.

"I think the field has a checkered past. I think a lot of historical concepts are frankly dreamed-up nonsense, but nonsense that can teach us principles that help move us along to a better place. I think the fact that therapies as varied as cognitive behavioral therapy and psychoanalysis are all shown to be helpful means that the common denominator, the human connection that they impose upon a patient, is the most important part."

She nodded along, and her eyes motioned for me to keep going.

"I think for some people the drugs are helpful. For some they're flat-out essential."

I thought of patients with severe mental illness, such as Ginger and Roger from my first overnight shift.

"Then there are those patients we just don't know how to help yet."

My mind went to Jane and then Elise.

"Like all fields, there are patients who suffer, and many have bad outcomes," she replied. "That doesn't mean we can't help them. In my experience, sometimes being a therapeutic partner for someone, even when you know the odds are against you, can be very rewarding for both the patient and the doctor."

• • •

I left the session recommitted to trying to break through to Elise. She might not get better, but she felt as though her parents, her friends, and even her school had abandoned her. I wouldn't do that to her.

When it came time for our next appointment, I refreshed the system obsessively to see if she had checked in. Our appointment time, 2 p.m., came and went. Then 2:02, 2:03, 2:04 passed as well. Finally, at 2:18, she arrived.

"I wasn't sure you were going to come," I said.

"Here I am," she replied.

After I led her into the office, she sat in my usual chair and dared me with her eyes to call her on it. I hadn't read anything about this kind of conflict in any of my textbooks, but without my conscious permission, a smile crept over my face.

"Why are you smiling?" she said with a sneer.

"Am I? I guess I'm just glad that you came back so we can get to work."

We met weekly for the rest of the year, and like Jane, she never stopped challenging me, but she did let me in to better understand how she was suffering. I wouldn't say she had achieved full remission by the time we finished in therapy together, but she had done well enough that I could see her as a complete person, no longer the overlapping set of diagnoses her chart had introduced to me months earlier. I had thought she was Jane Redux, and it sent me into a panic. Only after I gave her the space and opportunity to be who she was, and not who I saw her to be, did I create the room for her to change in the therapy. At our final session, following several weeks of preparing to end our work together and reflect on what we had accomplished, she thanked me for not ever giving up on her, and I said the same thing.

A couple of years after last seeing her, I received a letter in the mail. After our treatment ended, Elise had not only returned to school but graduated magna cum laude and was entering a graduate program in psychology.

35

This Is Where Change Happens

I sat in Meg Mook's office staring out the large windows behind her desk. I had only had the use of internal offices with no windows and plenty of bad artwork on the walls. As I approached the end of my time in residency, the thought of being hired at a place with windows made my chest feel warm with anticipation.

"So, here we are," Dr. Mook said.

"I wasn't sure we'd ever get here, but here we are."

It was the clinical exit interview all residents did to discuss how their clinical caseload was doing and who might be good candidates to receive patients staying on in the resident clinic for ongoing care.

We spent over an hour poring over my patient list as I tried to find the words to describe how these human beings had or hadn't progressed during our time together. I told Dr. Mook about the progress I had made with Elise, but also about having to close the chart on Oren, who had been lost to follow-up.

"What about Jim?"

Jim was my first therapy patient, and I felt a particular sadness about how I was leaving him.

"Jim is still Jim," I said.

She looked at me, puzzled.

"He's still a raging narcissist — his words, not mine — but now his wife has left him and he's finding new women to temporarily fill the void of his fractured self."

She looked at me with eyebrows raised.

"Geez, I sound like a real Harvard shrink."

"I think you might be one," she replied. "Speaking of which, have you started looking for jobs yet? For after graduation, I mean. You know, we'd love to find a position here for you."

My mouth opened to answer, but nothing came out.

"Don't tell me now. Think it over."

The truth is that I had already thought it over a great deal. I had even interviewed around town. There was a large group practice north of Boston that ran like a well-oiled machine with twenty-minute appointments and twenty-five scheduled patients per day at one of several locations. This place felt like the kind of doctor's office you'd find in Beverly Hills, with glitzy sculptures and artwork tastefully placed on the walls. The starting salary was substantial, but I couldn't see myself finding satisfaction in such brief interactions with such large caseloads. I was doubtful I'd get to know any patients very well at all in such a setting.

The next place I interviewed was a lovely practice with three locations around Boston proper. It had a very different vibe because there was only one psychiatrist at each location and dozens of therapists and neuropsychologists. Again, the locations seemed wonderful — much glitzier than the offices at Longwood, but I couldn't get beyond the idea of being asked to focus primarily on the med-management side of patient care, while the therapists around me got to better know their patients.

I looked at a group practice south of Boston that I adored. It comprised MDs, who shared resources but practiced however they wanted as individual clinicians. Some earned revenue went back to the practice for overhead and administrative staff, while the remaining take-home entirely depended upon how I would choose to practice. It had a lot of appeal, and I was close to signing with them.

The factor that held me back was the disconnect from the academic community I had come to consider my home. There would be no grand-rounds discussions of emerging research topics, no morbidity and mortality conferences to attend to learn from one another's mistakes. Perhaps they made up for it informally in the break room, but I wouldn't have that essential responsibility to students and residents that I would if I stayed in academia. Teaching had become one of the most revitalizing parts of the job for me in that fourth year, and I was hesitant to give it up.

I circled back to my own hospital. I continued to find the TMS work enthralling — particularly when paired with therapy — and so I asked the leaders of the TMS program if they ever thought about bringing on a psychiatrist full-time to work with them. They were delighted with my proposal, but they didn't want to hire me as a simple bridge to psychiatry. They wanted me to become a member of their team and help advance the field.

"This is where change happens," one said. "It's an amazing place to come to work every day."

The pay would be lower than I could get elsewhere, and initially I'd have to share an office again — maybe with a window, and maybe without — and I'd probably have to fill in my hours with other kinds of clinical work, but I'd be able to get in on the ground floor of a treatment approach that was just starting to take off. I could become involved in research projects that would pave the way for new treatments for patients when no other treatment worked for them. Best of all, I could stay within the community of Harvard Medical School and become an instructor, taking over the first-year psychopharm course and supervising residents, helping them to develop over their four years. I took the job and went back to Dr. Mook to let her know I'd be staying but working mainly on the other side of the building. She was very happy for me and reminded me that her door was always open — a perk I've taken advantage of many times in the years since, whenever I find myself in a precarious clinical situation. She also alerted me to a policy that allowed faculty to rent space in the evening hours to see patients privately. The rates were exceedingly competitive, and I might even be able to splurge for an office with a window.

Our residency graduation took place in the middle of June. My family came up from New York for the event and schmoozed with the faculty during a sort of reception before the official ceremony. One by one, my mentors came over and told my parents what a privilege it had been to guide me over the previous four years. There were Nina and Jen, Strand, McQueen, Mook, Redding, and Macy all circling, singing my praises. I couldn't understand it at the time. I knew with certainty how many times throughout residency I had been mediocre, insufficient, or even just too exhausted to engage. I wondered how these people could feel so proud to mentor someone who had been so helpless for so long.

Years later, though, I began to understand. On the faculty, we watch residents come into the program wide-eyed and eager but lacking the scars from loss that come with clinical experience. To help someone through that process and watch her become a fully formed, if less perfect and more battle-worn, version of herself is a gift that few get to experience.

During the official ceremony, each resident was celebrated in a toast by a faculty member of their choosing. Dr. Mook spoke for me, and what I remember most is how thrilled she was that I was staying close and remaining a part of our community. Finally, to cap off the evening, the head of the department awarded me with the Henry G. Altman Award for Excellence in Medical Education. He said that this year's class had so many outstanding residents that it was difficult for the committee to choose just one, but that I had earned this special recognition for my work with the medical students and junior residents. He shook my hand, and I wondered if *this* is what "making it" feels like. Years later, when finally I did have an office of my own large enough to display diplomas on the wall, I hung up that award right next to them. It remains my most prized accolade.

Two weeks after graduation, Rachel and I got married in a small ceremony in the Boston Public Garden surrounded by our closest friends and family, including several members of our Golden Class. Dozens of onlookers in the public park stopped what they were doing to watch. It was impossible not to appreciate the irony of celebrating our love so publicly, and smile.

During the recitation of the vows, our eyes were full and our voices quivered. The ceremony paused, not once but twice, as a vibrantly dressed man with a loud eighties-style boombox doing slow laps around the garden sauntered by playing music that sounded vaguely Mediterranean. Our eyes connected, and we smiled at each other, sharing a moment of pure levity. A few moments later I slipped Rachel's perfect ring on and held her hand. When the brief ceremony was complete, we exhaled together, appreciating what we had become together. The rest of our lives awaited.

We put off honeymooning until the next time we could scrape together two weeks off at the same time, which wouldn't be until September. Two days after marrying Rachel, I returned to Longwood to begin my new faculty job. I knew I would never be the version of the mythical Harvard psychiatrist that had existed in my mind four years earlier. I had seen too many examples of shared humanity among the patients and those trying to help them to be hung up on formalities. The space where that psychiatrist had once existed in my mind had been filled instead with hard-earned truths about what it means to connect to those people around you, to commit to them, and to purposefully keep moving forward.

In Memoriam

Three years after graduating and dispersing around the country to begin the next phase of our lives, our class came together again in Boston to grieve the loss of one of our own. Dr. Christine Petrich was a beloved member of our group and a good friend to me during our four years together. After we graduated from Harvard Longwood, we had largely lost touch as she went on to do a prestigious fellowship in forensic psychiatry and then to take an academic faculty position close to family in Texas. One morning we all woke up to the news that she had ended her own life. Collectively, we were shocked and heartbroken. We each tried to make sense of her suicide in our own way but felt that we couldn't fully grieve her loss unless we processed it together. Nina hosted us all at her house. Jen came, too, and the few residents who were too far to join us attended by phone. We held our final Feelings class that night over dinner. We shared stories and expressions of the grief we felt. The world had lost someone very special.

I wondered aloud if there wasn't more that I could have done for Christine, and others said they struggled with the same doubts and regrets. We were supposed to be experts in preventing suicide, but we hadn't protected one of our own. We pushed aside the temptation to dissect our time together for missed signs of danger. Whatever risk factors we could retrofit onto our friend's life wouldn't bring us closer to resolving our grief, and the fact is that each one would be overmatched by the wonderful, invaluable aspects of her life—her intellect, achievement, and love for her family and friends. Christine could have been any one of us, and her loss brought the preciousness of our own lives into greater focus.

By the end of the evening, we had shed tears for our friend and celebrated the special person she was. We resolved to be there for one another in the years ahead without hesitation or reservation. As we said our goodbyes once again, we embraced more tightly and felt more at peace than when we had begun.

Acknowledgments

So many people supported me in the journey of writing this book. I'd like to begin by thanking my classmates at Harvard Longwood. My story of becoming a psychiatrist is inseparable from theirs. I appreciate their encouragement to go forward with the publication of our shared experience. Their collective memory and input on particular content was also invaluable. To my former colleagues and lifelong friends from this group, thank you.

Thanks also to the teachers I have had within psychiatry and throughout my entire life. At Harvard Longwood alone there were countless individuals dedicated to teaching our class what it truly means to be a compassionate and competent psychiatrist. I have failed to adequately honor many of them specifically in these pages, but I thank them all, nonetheless.

These lessons could certainly not have been learned without the patients I saw. I hope that my story conveys my incalculable appreciation for our shared humanity and the gratitude I feel for the opportunity to work with them. The practice of medicine calls for an unending drive to do better by our patients, and one of my promises to patients — past, present, and future — is that I will continue to strive to be a better partner in care.

I'd like to express more broad and everlasting gratitude and love to my family. For years they have led me to believe that I could accomplish great things in this life, and without that frame, I could never have achieved the life I've made. To my parents, Terrie and Mark, I wouldn't be a writer and a doctor without your genes, and I wouldn't be any good at either one without your unconditional love and care. You have given me a drive to do good in the world and be a source of light in dark places. Thank you. To my brother, David, you have always been my guide and role model. When I achieve my dreams, it is often because you have shown me the way. This was true on the road to medicine and remains true in life. My uncles Steven and Jeffrey — you have both taught me that for people like us creativity can be at the very core of a person throughout his entire life and that this trait is to be embraced. In childhood, it came naturally to me, but I learned from you how to cultivate my creativity and let it breathe throughout adulthood. Thanks also to my cousin Brad for expressing amazement and encouragement at even my most trivial medical accomplishments and to all of my aunts, uncles, cousins, grandparents, in-laws, nieces, and nephew — I love you all and appreciate the love I have felt back from you. You have supported me so thoroughly in this life, and I couldn't be more grateful. To my sons I want to convey that, like your mother, you have enriched my life in unimaginable ways. My dream for you is to live a life as full of wonder, joy, and curiosity as you bring to the world. I love you both more than words can express.

I would like to thank my literary agents, first Karen Murgolo and now Michael Signorelli at Aevitas Creative Management. Karen, you took a chance on me and helped guide me through the process. I can say with certainty that this book would not exist without you. Michael, you have helped move it forward to completion, affording me the opportunity to achieve a lifelong dream. Similarly, I want to extend my enormous thanks to Deb Brody and the entire team at HMH Books. I have known from the very beginning that my story was in the best possible hands with you.

Infinite gratitude also goes to the medical team that has kept me alive and well long enough to write this book and, more important, to reach for ever higher spoils in this life. Particular thanks go to Toni Choueiri, Jason Petrilla, and the teams at Dana Farber Cancer Institute, Brigham and Women's Hospital, and Beth Israel Deaconess Medical Center. Appreciation also goes to my work colleagues and many friends, who have supported me unconditionally through the last few challenging years.

Thank you also to Silvia Graziano, Michael Kleiman, Roscoe Brady, Erica Greenberg, and Liz Perez-Daple for your editorial guidance and friendship. Voices like yours in the earliest phases of this book's creation were essential to making it happen, and I will be forever grateful. I would also like to offer special thanks to Alan Barnett, Michaela Labriole, and Deirdre Neilen, three people from different corners of my world who made me believe in myself as a writer.

Finally, and most essentially, I would like to thank my wife for allowing me to tell my particular version of our story. As is clearly evident, she is a woman who appreciates a certain degree of personal privacy, and telling the story of how we came together during residency probably feels to her something akin to the flooding technique in exposure therapy. Still, she supported me in the project, and that means the world to me. She is a remarkable person whose distinct charm and grace cannot be adequately transcribed onto the page. While I did the best I could to convey how I fell in love with my wife, I know that the result does not nearly do justice to the complexity and richness of our relationship or to her as a beautiful, complete person.